Advances in Pattern Recognition
and Artificial Intelligence

Series on Language Processing, Pattern Recognition, and Intelligent Systems

Print ISSN: 2661-4316
Online ISSN: 2661-4324

Co-Editors

Ching Y. Suen
Concordia University, Canada
parmidir@enes.concordia.ca

Lu Qin
The Hong Kong Polytechnic University, Hong Kong
csluqin@comp.polyu.edu.hk

Published

Vol. 1   Digital Fonts and Reading
         edited by Mary C. Dyson and Ching Y. Suen

Vol. 2   Advances in Chinese Document and Text Processing
         edited by Cheng-Lin Liu and Yue Lu

Vol. 3   Social Media Content Analysis:
         Natural Language Processing and Beyond
         edited by Kam-Fai Wong, Wei Gao, Wenjie Li and Ruifeng Xu

Vol. 4   Computational Linguistics, Speech and Image Processing for
         Arabic Language
         edited by Neamat El Gayar and Ching Y. Suen

Vol. 5   Frontiers in Pattern Recognition and Artificial Intelligence
         edited by Marleah Blom, Nicola Nobile and Ching Y. Suen

Vol. 6   Advances in Pattern Recognition and Artificial Intelligence
         edited by Nicola Nobile, Marleah Blom and Ching Y. Suen

Series on Language Processing, Pattern Recognition,
and Intelligent Systems — Vol. 6

Advances in Pattern Recognition and Artificial Intelligence

Edited by

Nicola Nobile
Marleah Blom
Ching Y. Suen

Concordia University, Canada

World Scientific

NEW JERSEY · LONDON · SINGAPORE · BEIJING · SHANGHAI · HONG KONG · TAIPEI · CHENNAI · TOKYO

Published by

World Scientific Publishing Co. Pte. Ltd.

5 Toh Tuck Link, Singapore 596224

USA office: 27 Warren Street, Suite 401-402, Hackensack, NJ 07601

UK office: 57 Shelton Street, Covent Garden, London WC2H 9HE

British Library Cataloguing-in-Publication Data
A catalogue record for this book is available from the British Library.

Series on Language Processing, Pattern Recognition, and Intelligent Systems — Vol. 6
ADVANCES IN PATTERN RECOGNITION AND ARTIFICIAL INTELLIGENCE

Copyright © 2022 by World Scientific Publishing Co. Pte. Ltd.

For photocopying of material in this volume, please pay a copying fee through the Copyright Clearance Center, Inc., 222 Rosewood Drive, Danvers, MA 01923, USA. In this case permission to photocopy is not required from the publisher.

ISBN 978-981-123-900-7 (hardcover)
ISBN 978-981-123-901-4 (ebook for institutions)
ISBN 978-981-123-902-1 (ebook for individuals)

For any available supplementary material, please visit
https://www.worldscientific.com/worldscibooks/10.1142/12336#t=suppl

Preface

Following the success of the last edition, *Advances in Pattern Recognition and Artificial Intelligence* provides a collection of original works covering a range of topics within the area of pattern recognition and artificial intelligence, this volume brings together a new collection of up-to-date chapters written by a diverse range of international scholars.

Works were carefully selected from 100 papers submitted to the 2020 edition of the *International Conference on Pattern Recognition and Artificial Intelligence* (ICPRAI 2020). The conference, sponsored by the Centre for Pattern Recognition and Machine Intelligence (CENPARMI) as well as other scientific, technological and industrial partners, brought individuals from all over the world together through a lucrative virtual online conference. Fourteen papers presented at the international event were chosen and extended to showcase current and novel works produced by a diverse international group of authors.

Chapters cover a range of topics and applications, which highlight the many ways pattern recognition and artificial intelligence can be studied and applied in our complex and quickly changing world. In Chapter 1, Rouhafzay et al. propose, evaluate different strategies and provide a logical understanding for lesion detection in breast MRIs using deep learning models. Al-Qudah and Suen (Chapter 2) summarize recent contributions, discuss challenges, and outline future directions for computer-based analyses that use deep learning neural networks to assess health through blood smears. Security related practical applications are explored by Qin and El Yacoubi in Chapter 3. With the development of new and Generative Adversarial Network model to recognize and extract palm veins, their experimental results demonstrate the effectiveness of their proposed model, which has important implications for automatic biometric personal verification and information security.

In Chapter 4, Xu et al. then introduce a training model that uses deep learning, which improves the diagnostic accuracy of retinal disease. Chapters 5, 6, and 7 showcase interesting works within the area of handwriting recognition. Lowe (Chapter 5) addresses if cursive handwriting remains essential within the digital age as well as describes how patterns emerge and intersect into handwriting features that may be used to recognize individuals' personality traits. A novel attempt to investigate handwriting difficulties through the lens of machine learning and pattern analysis is then described by Adak et al. in Chapter 6. The authors analyze one individual's handwriting to investigate challenges writing character shapes before describing their proposed deep reinforcement learning-based model to predict difficulty measures. In Chapter 7, Al-Qawasmeh and Suen then outline a new way to detect the writer gender from scanned handwritten documents based on the concept of transfer-learning.

The remaining chapters address topics including image analysis, financial applications, database creation, text analysis, as well as neural networks. Chapter 8 presents a descriptive overview of the Wartegg Test, a semi structured drawing test used assess personality characteristics. Within the chapter, Pettinati notes the use of the test in a range of context as well as discusses possibilities and challenges of scoring, analyzing and interpreting results with the help of artificial intelligence. Zhang et al. (Chapter 9) provide a new credit scoring model after which experiments resulted in a high prediction accuracy to assist lenders when making prediction models for credit scoring. Vincent and Kurtz (Chapter 10) highlight diversity and properties of features when describing optimal choices for designing image analysis systems. Nobile et al. (Chapter 11) describe a comprehensive and unconstrained license plate database that includes a varied and novel collection of images of license plates. Vidal et al. propose an approach based on neural models to classify Spanish criminal news in Chapter 12. Chan et al. (Chapter 13) train a Deep Convolutional Neural Network, apply the model using Twitter then compare their predicted scores to actual results from the 2016

Presidential Election in the United States. Lastly, a new two-step feature algorithm is proposed with a later joint version being introduced by Tan et al. in Chapter 14. The authors also design two interactive algorithms and demonstrate their effectiveness.

Nicola Nobile, Marleah Blom and Ching Y. Suen

Contents

Preface.. v

List of Contributors... xi

Chapter 1: Enhanced Lesion Detection in Breast MRI Using Parallel 1
and Cascaded Integration of Deep Learning Models
Ghazal Rouhafzay, Yonggang Li, Haitao Guan, Chang Shu,
Rafik A.-Goubran and Pengcheng Xi

Chapter 2: A Survey on Peripheral Blood Smear Analysis Using.......... 23
Deep Learning
Rabiah Alqudah and Ching Y. Suen

Chapter 3: End-to-End Generative Adversarial Network for.................. 47
Hand-Vein Recognition
Huafeng Qin and Mounim A. El Yacoubi

Chapter 4: Diabetic Retinopathy Analysis Based on Retinal Fundus 61
Photographs via Deep Learning
Yan Xu, Yufang Tang, Xuezhou Wen and Ching Y. Suen

Chapter 5: Pattern Recognition in Handwriting..................................... 77
Sheila Lowe

Chapter 6: A Deep Reinforcement Learning-based Study on................. 97
Handwriting Difficulty Analysis
Chandranath Adak, Bidyut B. Chaudhuri and Michael Blumenstein

Chapter 7: Gender Detection from Handwritten Documents................. 119
Using the Concept of Transfer Learning
Najla AL-Qawasmeh and Ching Y. Suen

Chapter 8: The Wartegg Test 133
Graziella Pettinati

Chapter 9: A New Prediction Method for Credit Scoring Based on 145
Sampling Reconstruction of Signal on Graph
Qian Zhang, Zhihua Yang, Feng Zhou and Lihua Yang

Chapter 10: Optimal Choices of Features in Image Analysis 161
Camille Kurtz and Nicole Vincent

Chapter 11: A Comprehensive Unconstrained, License Plate 183
Database
Nicola Nobile, Hoi Kei Phoebe Chan and Marleah Blom

Chapter 12: Classification of Spanish Criminal News using 193
Neural Networks
Mireya Tovar Vidal, Emmanuel Santos Rodríguez and
José A. Reyes-Ortiz

Chapter 13: Predicting US Elections with Social Media and 213
Neural Networks
Ellison Yin Nang Chan, Adam Krzyżak and Ching Y. Suen

Chapter 14: Differences and Similarities learning for Unsupervised ... 231
Feature Selection
Tan Jun, Xinyi Li, Ning Bi and You Ning

Index 259

List of Contributors

Rafik A.-Goubran
Carleton University, Ottawa, Canada
goubran@sce.carleton.ca

Chandranath Adak
JIS Institute of Advanced Studies and Research, JIS University, Kolkata, India
adak32@gmail.com

Najla Al-Qawasmeh
Concordia University - CENPARMI, Montréal, Canada
n_alqawa@encs.concordia.ca

Rabiah Al-Qudah
Concordia University - CENPARMI, Montréal, Canada
r_alquda@encs.concordia.ca

Ning Bi
Sun Yat-Sen University, Guangzhou, China
mcsbn@mail.sysu.edu.cn

Marleah Blom
Centre for Pattern Recognition and Machine Intelligence
Concordia University, Montréal, Canada
marleah.blom@concordia.ca

Michael Blumenstein
University of Technology Sydney, Sydney, Australia
michael.blumenstein@uts.edu.au

Ellison Chan
Concordia University, Montréal, Canada
ellison.chan@gmail.com

Hoi Kei Phoebe Chan
Concordia University - CENPARMI, Montréal, Canada
hkpc07@gmail.com

Bidyut B. Chaudhuri
Techno India University, Kolkata, India
bbcisical@gmail.com

Mounim A. El-Yacoubi
Telecom SudParis, Évry, France
mounim.el_yacoubi@telecom-sudparis.eu

Haitao Guan
Nantong No. 3 People's Hospital, China, Nantong, China
happyght@qq.com

Tan Jun
Sun Yat-Sen University, Guangzhou, China
mcstj@mail.sysu.edu.cn

Adam Krzyzak
Concordia University, Montréal, Canada
krzyzak@cs.concordia.ca

Camille Kurtz
Université de Paris, Paris, France
camille.kurtz@u-paris.fr

Xinyi Li
Sun Yat-Sen University, Department of Mathmatics
Guangzhou, China
390860549@qq.com

Yonggang Li
First Affiliated Hospital of Soochow University, China, Suzhou City, China
liyonggang224@163.com

Sheila Lowe
Sheila Lowe Handwriting Examiner, United States
sheila@sheilalowe.com

You Ning
Sun Yat-sen University, Guangzhou, China
youn7@mail2.sysu.edu.cn

Nicola Nobile
Centre for Pattern Recognition and Machine Intelligence
Concordia University, Montréal, Canada
nicola@encs.concordia.ca

Graziella Pettinati
Graziella Pettinati Inc., Montréal, Canada
info@graziellapettinati.com

Huafeng Qin
Chongqing Technology and Business University, Chongqing City, China
qinhuafengfeng@163.com

José A. Reyes-Ortiz
Universidad Autónoma Metropolitana, Mexico City, Mexico
jaro@azc.uam.mx

Emmanuel Santos Rodriguez
Benemérita Universidad Autónoma de Puebla, Puebla, Mexico
e.ss.rdz@gmail.com

Ghazal Rouhafzay
National Research Council Canada, Ottawa, Canada
g.rouhafzay@gmail.com

Chang Shu
National Research Council Canada, Ottawa, Canada
Chang.Shu@nrc-cnrc.gc.ca

Ching Y. Suen
Centre for Pattern Recognition and Machine Intelligence
Concordia University, Montréal, Canada
suen@encs.concordia.ca

Yufang Tang
Shandong Normal University, Jinan, China
tangyufang@outlook.com

Mireya Tovar Vidal
Benemérita Universidad Autónoma de Puebla, Puebla, Mexico
mireyatovar@gmail.com

Nicole Vincent
Université de Paris, Paris, France
nicole.vincent@math-info.univ-paris5.fr

Xuezhou Wen
New York University, New York, U.S.A.
xw2447@nyu.edu

Pengcheng Xi
National Research Council Canada, Ottawa, Canada
perryxi@gmail.com

Yan Xu
Shandong Normal University, Jinan, China
yan.soe1011@gmail.com

Lihua Yang
School of Mathematics, Sun Yat-sen University; Guangdong Province Key Laboratory of Computational Science, Guangzhou, China
mcsylh@mail.sysu.edu.cn

Zhihua Yang
Information Science School, Guangdong University of Finance and Economics, Guangzhou, China
yangzh@gdufe.edu.cn

Qian Zhang
College of Mathematics and Statistics, Shenzhen University, Shenzhen, China
mazhangq@szu.edu.cn

Feng Zhou
Information Science School, Guangdong University of Finance and Economics, Guangzhou, China
fengzhou@gdufe.edu.cn

Chapter 1

Enhanced Lesion Detection in Breast MRI Using Parallel and Cascaded Integration of Deep Learning Models

Ghazal Rouhafzay[1, 2], Yonggang Li[4], Haitao Guan[5], Chang Shu[1],
Rafik Goubran[3], and Pengcheng Xi[1, 3]

[1] National Research Council Canada, Ottawa, ON K1A 0R6, Canada
[2] University of Ottawa, Ottawa, ON K1N 6N5, Canada
[3] Carleton University, Ottawa, ON K1S 5B6, Canada
[4] First Affiliated Hospital of Soochow University, Jiangsu, 215006, China
[5] Nantong No. 3 People's Hospital, Jiangsu, 226000, China

pengcheng.xi@nrc-cnrc.gc.ca; liyonggang224@163.com

Abstract: Breast cancer is the most common cancer in women worldwide. Computer Aided Detection (CADe) has attracted increasing research interest in recent years. Data exploration and lesion detection in medical images are tedious but can be accelerated using computational intelligence. One approach is to adapt and configure recent deep learning-based object detectors from computer vision to detect abnormalities in medical images. This chapter starts with three state-of-the-art object detectors, namely Faster R-CNN, YOLO v2 and Grad-CAM, to determine the location of lesions in Magnetic Resonance Images (MRI) of breast. Each individual detector is first tuned through adjusting network hyper-parameters and backbone architectures, in order to maximize Average Precision (AP). Two different integration approaches, namely cascaded and parallel integrations, are then proposed and implemented to improve the AP. The cascaded integration builds on a coarse-to-fine strategy. It uses Grad-CAM to compute coarse locations of bounding boxes for lesions and then applies YOLO v2 or Faster R-CNN as a fine detector. In the parallel integration approach, the detection result is a combination of results from the three detectors, with the aim of reducing missed detections. The integrated deep learning models exhibit enhanced performance on the detection of breast lesions in MRIs.

Keywords: Object detection, Lesion detection, Deep learning, Breast cancer, Magnetic resonance imaging

1   Introduction

As the most common cancer in women worldwide, breast cancer is receiving a huge research attention in terms of revealing influential factors and searching for diagnosis and treatment solutions. Early diagnosis of breast cancer is proven to play a pivotal role in survival rate. Despite the fact that mammography is the most effective tool for breast cancer screening, Magnetic Resonance Imaging (MRI) is recommended to assess the extent of breast cancer due to its capability of monitoring blood flow into tumors. Furthermore, breast MRIs can be effective where breast tissue is dense and mammogram or ultrasound (US) imaging fails to detect lesions; however, the false positive rate of tumor diagnosis using breast MRIs is reported to be relatively high, compared to other imaging modalities. This is due to the fact that MRI interpretation calls for a high level of expertise. A study conveys the specificity of human judgment on this imaging modality to be between 75% and 87% [1].

Computed Aided Detection (CADe) is an open field of research for assisting radiologists on tackling with challenging cases. In particular, the advancement in deep learning has made great strides in biomedical image processing. Once deep neural networks are properly trained, they are capable of capturing nonlinear and non-stationary features of training data. As a result, they can extract key features and subtle structures of medical images, which are difficult for human eyes to identify.

This work proposes and evaluates two different deep learning-based architectures through parallel and cascaded integration of state-of-the-art deep detectors in order to achieve the best performance. It also represents a framework leveraging global and local views for solving the problem. As part of the architecture, our approach provides explainable reasoning for lesion detection in breast MRI. Specifically, we focus on strengthening each method through analysing their performance to build an optimal decision-making strategy.

The chapter is organized as follows. Section 2 discusses relevant work from the literature. The integrated approaches are introduced in Section 3. Results are reported and discussed in Section 4. Section 5 concludes the chapter.

2   Literature Review

Complex nature of medical data calls for powerful tools being capable of extracting and learning biological patterns. Accordingly, it has received great interest in using deep Convolutional Neural Networks (CNN) for processing medical images. CNNs have been successfully applied to radiotherapy [2], attenuation correction for MRI and Positron Emission Tomography (PET) [3], tissues segmentation [4] and lesion detection [5], [6]. Moreover, the high rate of breast cancer in women worldwide has attracted enormous research attentions. As a result, machine learning solutions have been applied to different breast imaging modalities, including mammograms [5], ultrasound [6], microwave breast imaging [7] and MRI [8], [9], [10] to provide better treatment planning for this disease.

Among all the breast screening techniques, MRI provides decisive information about the nature of breast tissue and tumor characterizations; however, this imaging modality is prone to human error since it requires high level of expertise in interpretation. Ha et al. [8] designed a deep CNN network with ten convolutional layers to classify breast tumors detected in MRI to see if the tumor would respond completely, partially or not respond to chemotherapy during treatment. They succeeded in predicting up to 88% about the outcome of the treatment. In another research [9], the authors designed a deeper network with residual feedbacks to determine the subtype of tumors, which is a crucial step in treatment planning. Herent et al. [10] took advantage of a ResNet50 [11] network to extract features from breast MRIs. Extracted features were fed into an attention block for local predictions as well as into a logistic regression module for predicting possible lesion types. Authors in [12] compared the performance of VGG [13] network pre-trained on images from another domain and fine-tuned on breast MRIs and claimed the superiority of the latter. In a recent research, Whitney et al. [14] conducted a comprehensive study on the success of transfer learning on different imaging modalities on breast.

Besides the classification of lesions in breast tissue, determining their exact locations in an image can be performed using deep CNNs. Al-Masni et al. [15] applied a YOLO [16] object detector to mammograms for determining mass locations. Detected masses were then fed into a binary classifier to determine its malignancy. Chiao et al. [17] leveraged a Mask R-CNN [18] for tumor detections in breast sonograms. Faster R-CNN [19] was applied to digital breast tomo-synthesis to detect masses [20].

In this work, we apply three different object detectors, namely YOLO v2 [21], faster R-CNN [19] and Grad-CAM [22], to breast MRIs for lesion detection. In each detector, three different CNN architectures are experimented with and results are compared. Six different decision-making frameworks are proposed to boost the overall accuracy of lesion detectors. The YOLO v2 and faster R-CNN conduct local detections while the Grad-CAM provides a global view for the detections. Moreover, the Grad-CAM provides explainable reasoning as part of the solution.

3 Abnormality Detection Using Deep Learning

A variety of applications have benefitted from transfer learning using pre-trained deep CNNs by integrating different computational modules. Object detection is one of such applications targeting both locating and classifying objects in a scene. This framework can be leveraged for exploring a medical image to locate and classify abnormalities. This section discusses three state-of-the-art methods for object detection.

3.1 Faster R-CNN

Faster R-CNN is the second successor of Region-based CNN [23], where the Selective Search algorithm [24] is replaced by a Region Proposal Network (RPN) guiding the process of object detection and hence accelerating the performance. A CNN is firstly trained on full images whose last shared convolutional layer outputs a feature map. The RPN applies a sliding window of size 3×3 to the feature map.

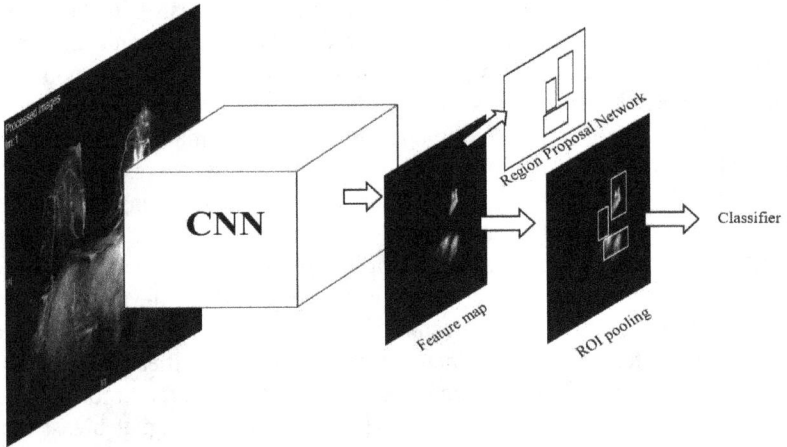

Fig. 1. Faster R-CNN for Lesion Detection.

The sliding window is convolved with the feature map to produce inputs for a classification as well as a regression network. It also accompanies 9 anchors as it passes the feature map and repeatedly measures the Intersection over Union (IoU) between each anchor and the ground-truth bounding box. The regression network is to learn an offset between the anchor and the ground-truth bounding box. If the IoU value is larger than 0.7, the anchor is positively weighted, otherwise if the IoU is less than 0.3, the anchor is negatively weighted in the loss function. Any anchor with IoU value in between is assigned a zero weight. Fig. 1 summarizes the overall framework of the Faster R-CNN for abnormality detection.

3.2   YOLO

YOLO is another deep learning-based object detector recognized for its real-time detection capability. It first divides a given image into an $S \times S$ grid where each grid is responsible for detecting an object whose center falls into that grid cell. The image passes through a deep CNN for learning a tensor of size $(S, S, B \times 5 + C)$, where C is the number of possible classes and B is the number of bounding boxes. The number 5 corresponds to the five parameters predicted for each cell of the grid, i.e. (x, y, w, h, P_c). x and y are the coordinates of the upper left corner of the bounding box and w and h are its width and height. P_c represents the confidence score of the bounding box. A threshold value (usually 0.25) is applied to remove boxes with low confidence values. The overall class confidence value is computed as $Confidence\ Score \times conditional\ class\ probability$. In the second version of YOLO, a series of refinements are applied to the model to increase the precision while reducing the running time. This includes adding batch normalization to convolutional layers, multiscale training, moving the class prediction from the cell level to bounding box level and picking better priors for anchors through a k-mean clustering algorithm. Instead of learning the offsets between the anchors and ground truth, the model learns the offsets with respect to the grid cells.

Fig. 2 illustrates the overall framework of YOLO. Training a YOLO detector requires training one single model and testing the model requires only one forward-pass computation. In comparison, the training of a Faster R-CNN model includes alternative steps, including training an RPN, reusing the convolutional part for training the detector, fine-tuning the RPN and re-training the detector.

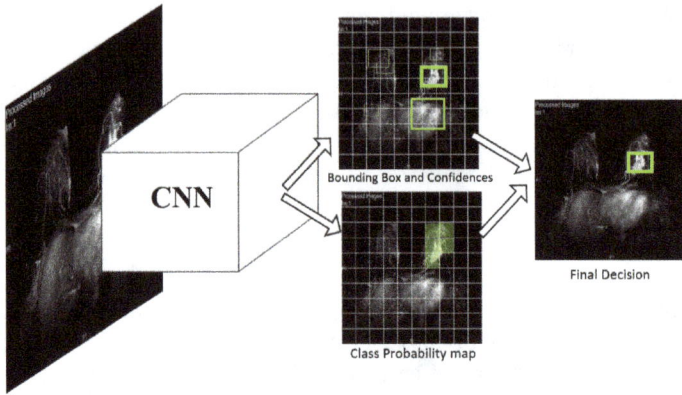

Fig. 2. YOLO v2 for Lesion Detection.

3.3   Grad-CAM

Using a Grad-CAM model for object detection starts from building a classifier. In this work, we start building the classifier using two sets of MRI images, one set labelled as normal and the other labelled as abnormal. Using transfer learning, these images are used to fine tune existing deep CNN networks in order to build a binary classifier (see Fig. 3). The output layer is then removed from the binary classifier in preparation for building a Grad-CAM model.

The idea of Grad-CAM is to compute the gradient of the final classification score with respect to the final convolutional layer in the network. Given a test image, if it is classified as abnormal by the binary classifier, the classification score will be used to compute gradients in regard to the last convolutional layer in the network. The gradients along with the convolutional feature maps are then linearly combined for computing a color map, which highlights the class-specific regions in the input image. The last step in this approach is to convert the abnormality map into a binary image and compute bounding boxes. They are then used to compare with ground-truth bounding boxes for verifications.

Fig. 3 introduces the abnormality detection approach using Grad-CAM in detail: first, two sets of training images are fed to a deep CNN for training a binary classifier, which is then configured for building a Grad-CAM model. Given a test image, if it is classified by the binary classifier as "abnormal", it will be fed to the Grad-CAM model for computing a color-coded map highlighting the abnormalities. The abnormality map is then converted to a binary map using a threshold value (=0.9), which is then used to compute a bounding box as the final detection.

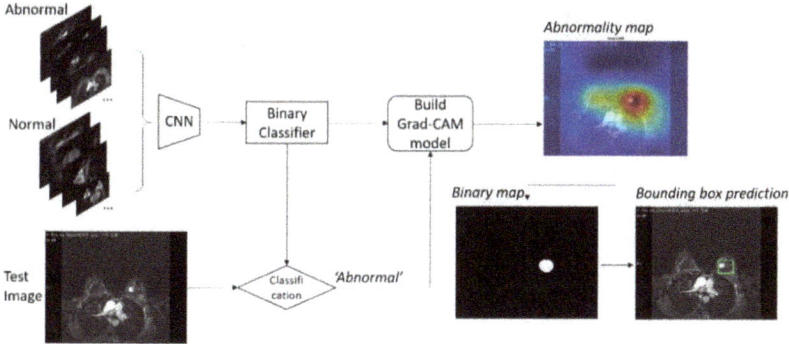

Fig. 3. Grad-CAM for Lesion Detection.

4   Integrated Approaches for Abnormality Detection

This section discusses a variety of integration approaches for us to make optimal predictions about the abnormal regions within an MRI. These approaches can be categorized into cascaded and parallel integrations. The cascaded integration includes applying two abnormality detectors in a successive way. In the parallel integration, we determine the abnormality region based on the outcome from different detectors.

4.1   Cascaded Integration

One possible approach to integrate the previously introduced abnormality detectors, is to apply them successively to an MRI image. Since Grad-CAM highlights the regions of an MRI where abnormality exists, it can be used to make a coarse estimation about the abnormal region in a full MRI. YOLO v2 and Faster R-CNN can then be applied to the coarse region of interest to make a fine decision about the bounding box delineating the abnormal region. Fig. 4 illustrates the framework of cascaded integration of the detectors. It is worth noting that the YOLO v2 and the Faster R-CNN detectors used in the second step of cascaded integration are trained on the cropped version of MRIs using the Grad-CAM coarse detector. In other words, we have applied the configured Grad-CAM to the training portion of abnormal images, and for each training image, we computed a normalized color-coded map. Training images are then cropped around a binary map with a threshold value of 0.6. Bounding boxes for the training set are then recomputed for the cropped region and a new YOLO v2/Faster R-CNN network is trained on cropped images. In the testing phase, full MRI images of the test set are firstly fed into the binary classifier. A coarse detection region is cropped and then fed into YOLO v2/Faster R-CNN networks. Detected bounding boxes on cropped images are transformed back to the original image for evaluation purposes.

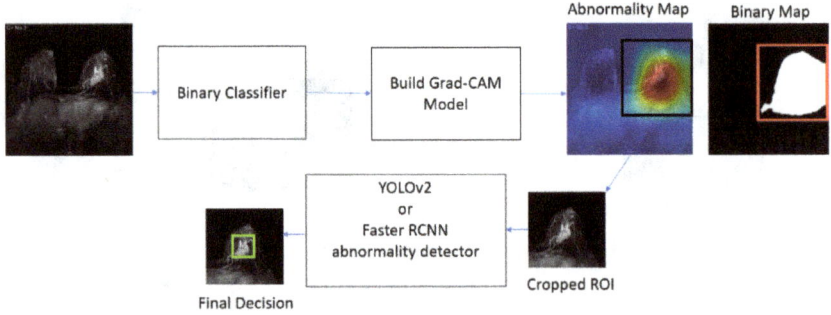

Fig. 4. Cascaded Integration of Abnormality Detectors.

4.2 Parallel Integration

Six different parallel integration approaches are proposed in this section through combining final results from the implemented lesion detectors to boost the overall performance of lesion detection in terms of decreasing the missing rate and enhancing the Average Precision (AP). The overall integration method (referred to as all methods in Section 5) is formulated as follows to minimize the number of missed detections, capitalizing on results from the three detectors by combining the results from YOLO v2 and Faster R-CNN detectors. To achieve this, for each test sample the detected bounding boxes by Faster R-CNN and YOLO v2 are computed and the overlapping ratio between all detected bounding boxes is computed. If the overlapping ratio is more than 50%, the two detections are considered to be identical. In that case, a linear regression model is used to determine the final bounding box of such abnormality. If the overlapping ratio is less than 50%, the integrated solution takes in both detections as a positive case. Subsequently, the detected bounding boxes are compared with the boxes achieved by Grad-CAM method and the overlapping ratios are computed. Again, if the overlapping ratio between the detection by Grad-CAM and YOLO v2/Faster R-CNN is less than 50%, the Grad-Cam detection is added as a new detection case. The overlap ratio is computed as follows:

$$\text{overlap ratio} = \frac{\text{Faster R--CNN bounding box} \cap \text{YOLO v2 bounding box}}{\min(\text{Faster R--CNN bounding box}, \text{YOLO v2 bounding box})} \quad (1)$$

Here is the pseudocode summarizing the proposed approach:

```
for bounding boxes detected by YOLO v2 and Faster R-CNN
   Compute the overlap ratio
   if overlap ratio ≥ 0.5
      Train four linear regression models to learn a mapping
   of the four parameters of the predicted bounding boxes
   to the ground truth.
   elseif overlap ratio < 0.5
      Add all detected bounding boxes
for bounding boxes detected in the previous step and by
Grad-CAM
   Compute the overlap ratio
   if overlap ratio < 0.5
      Add the detection by Grad-CAM to the results
```

We use Regression Learner App in Matlab (MathWorks, USA) and train the regression models on bounding boxes detected by Faster R-CNN and YOLO v2 object detectors. Four such linear regression models are trained to predict the weights of the bounding boxes (i.e. the coordinates of the upper left corner as well as the width and the length). Results of this solution are presented in Section 5 as "All Methods". The other five integration strategies are defined as follows:

- **Faster R-CNN ∪ YOLO v2:** This method excludes the Grad-CAM detection from the overall integration method.
- **Faster R-CNN ∩ YOLO v2:** This method only includes linear regression output of matched detections between Faster R-CNN and YOLO v2.
- **Faster R-CNN priority:** This method includes linear regression output of matched detections between Faster R-CNN and YOLO v2 as well as detections by Faster R-CNN which are not detected by YOLO v2.
- **YOLO v2 priority:** This method includes linear regression output of matched detections between Faster R-CNN and YOLO v2 as well as detections by YOLO v2 which are not detected by Faster R-CNN.
- **Faster R-CNN ∩ YOLO v2 + Grad-CAM:** This method includes linear regression output of matched detections between Faster R-CNN and YOLO v2 as well as detections achieved by Grad-CAM.

5   Results and Discussions

Our dataset is composed of 3,161 MR images acquired by multiple MRI scanners. Images are of various sizes and taken from different viewpoints. We apply a data cleaning procedure to crop images with large margins in order to focus on the imaged body. 1,988 images of the dataset contain breast abnormality while 1,173 images are normal. The Faster R-CNN and YOLO v2 abnormality detectors are trained on the abnormal images. In the case of Grad-CAM, which requires a trained binary classifier for abnormal and normal classes, a balanced dataset is generated by random sampling. All networks are trained on a machine equipped with NVIDIA RTX 2070 graphics card with a split of 80/20 for training and testing. The available GPU memory supports training YOLO v2 models with a mini-batch size up to 32. While for faster R-CNN, the available GPU memory can only afford to train on a single image in a batch. Learning rate is also tuned for all networks for achieving the best performance.

Table 1 compares the AP values for YOLO v2, Faster R-CNN and Grad-CAM detectors with Resnet50, GoogleNet or MobileNet v2 as the backbone. The threshold value used in Table 1 is the ratio of Intersection over Union (IoU) between detected bounding box and ground truth for the detection being considered as successful. One can notice that for a threshold of 0.5, MobileNetV2 achieves the best AP compared to other architectures in YOLO v2. In Faster R-CNN, Resnet50 outperforms other architectures. For all backbone architectures, YOLO v2 achieves a higher AP compared to Faster R-CNN. Grad-CAM results in a lower AP since it does not directly rely on ground truth bounding boxes during training. Furthermore, this method looks at the image in a global scale. Nonetheless, when the IoU threshold is reduced, Grad-CAM starts to have detection capabilities, confirming the idea that the color mapping does point toward the location of abnormalities.

Table 1. Average Precision for Different Methods

Method	Resnet50		GoogLeNet		MobileNetV2	
	Th=0.1	Th=0.5	Th=0.1	Th=0.5	Th=0.1	Th=0.5
YOLO v2	0.8056	0.7431	0.6712	0.6232	**0.8228**	**0.7651**
Faster R-CNN	**0.8131**	**0.6993**	0.7663	0.6172	0.7859	0.6521
Grad-CAM	**0.2383**	**0.003**	0.0684	0.0000907	0.0407	0.000057

Fig. 5 compares the Precision Recall curves of YOLO v2, Faster R-CNN and Grad-CAM. One can find that, YOLO v2 with MobileNetv2 or Resnet50 backbone architectures have the best performance.

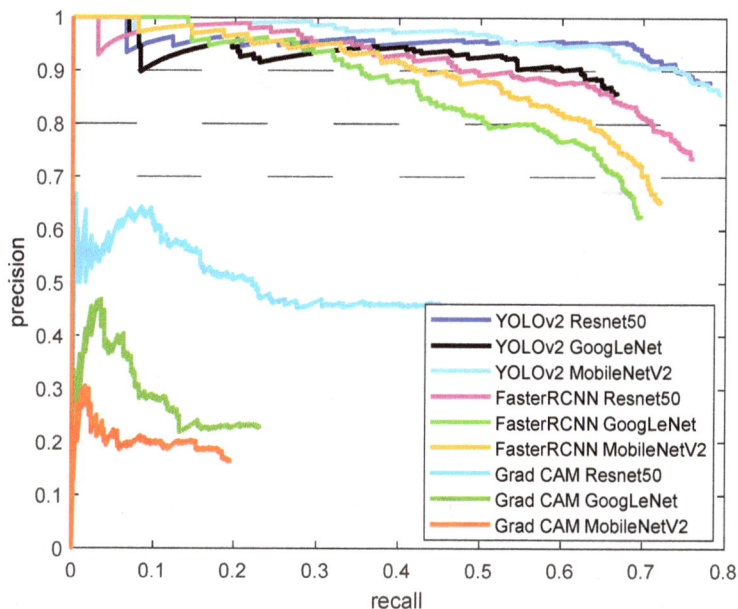

Fig. 5. Precision-Recall Curves for Different Abnormality Detectors.

When training an architecture for lesion detection, a crucial goal is to minimize the number of missed lesions. Therefore, we plot the Miss Rate against the False Positive per Image (FPPI) to find out which method can achieve the minimum Miss Rate. Evidently, allowing more false positive detections can reduce the number of missed detections. Fig. 6 compares the missed abnormality rates against FPPI for YOLO v2, Faster R-CNN and Grad-CAM detectors with different CNN architectures. This graph confirms that the minimum missed rate can be achieved by YOLO v2 abnormality detector with MobileNetV2 on the backbone.

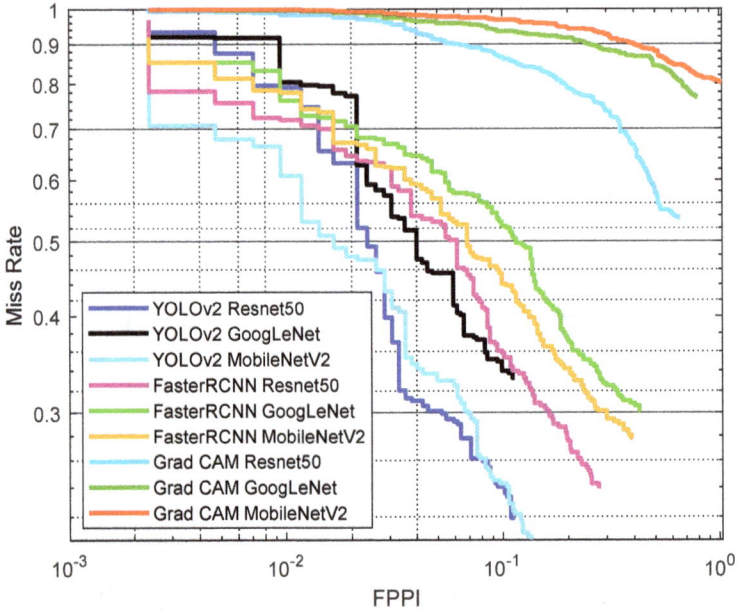

Fig. 6. Miss Rate Versus False Positive Per Image for Different Abnormality Detectors.

Table 2 provides the AP values for cascaded integration of Grad-CAM as a coarse detector followed by YOLO v2 or Faster R-CNN as a fine detector. As discussed in Section 4.1, the Grad-CAM framework is firstly applied to training dataset to achieve a coarse detection of abnormality. The latter forms the input to train new YOLO v2 and Faster R-CNN models for abnormality detection. Cascaded integration of Grad-CAM (th=0.6) and YOLO v2 can increase the AP of YOLO v2 object detector by 0.0236 while this improvement is about 0.0115. It is noted that the Grad-CAM detection as an individual detector employs a threshold of 0.9 to narrow down the decision bounding box to only highest values on the color-coded map. This value is reduced to 0.6 for Grad-CAM in the cascaded architecture to reduce the miss rate after Grad-CAM.

Fig. 7 and Fig. 8 compare the precision-recall and miss rate-FPPI curves for the cascaded integrated solutions. Both graphs confirm the superiority of the approach with YOLO v2 as fine detector.

Table 2. Average Precisions for the Cascaded Integration Strategies

	Grad-CAM + YOLO v2	Grad-CAM + Faster R-CNN
Average Precision	0.7887	0.7108

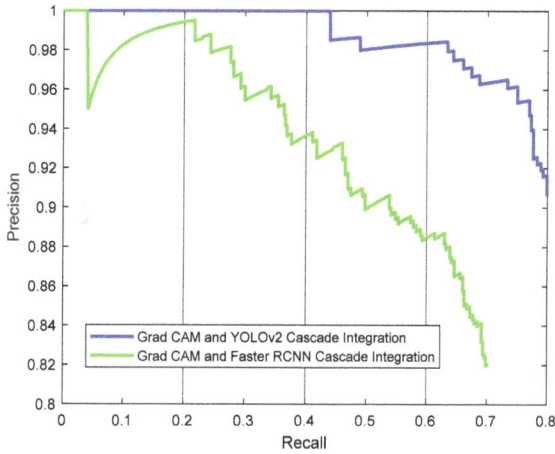

Fig. 7. Precision Recall Curves for Cascaded Integrated Solution.

Table 3 reports the AP values achieved with different parallel integration approaches. The Faster R-CNN ∩ YOLO v2 + Grad-CAM described in Section 4.2 achieves the highest AP value.

Table 3. Average Precisions for the six parallel integration strategies

	All Methods	Faster R-CNN ∪ YOLO v2	Faster R-CNN Priority	YOLO v2 Priority	Faster R-CNN ∩ YOLO v2	Faster R-CNN ∩ YOLO v2 + Grad-CAM
Average Precision	0.7259	0.6717	0.7387	0.7987	0.7586	**0.8097**

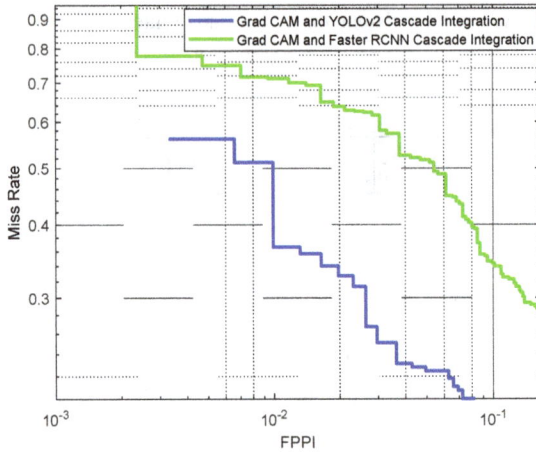

Fig. 8. Miss Rate Versus False Positive Per Image for Cascaded Integrated Abnormality Detectors.

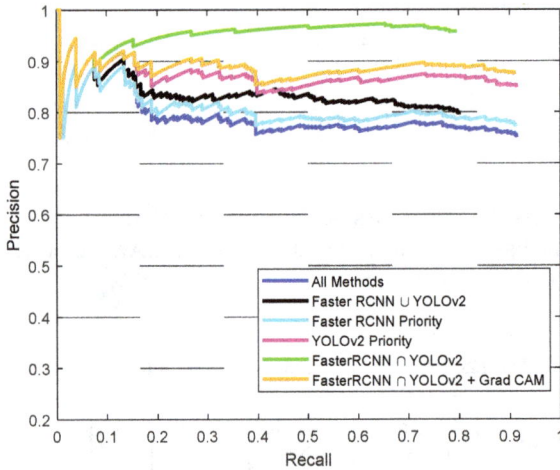

Fig. 9. Precision Recall Curves for Different Parallel Integration Methods.

Fig. 9 and Fig. 10 depict the precision-recall as well as the miss rate-false positive per image curves for different versions of parallel integration methods. The precision-recall curves suggest that the best performance can be achieved by computing the linear regression between the matched detections acquired by Faster R-CNN and YOLO v2. Nevertheless, one can notice that the approach relying on all detectors (all methods) yields the minimum possible missed detections. In other words, the Faster R − CNN ∩ YOLO v2 detector has the highest accuracy for lower recall values; however, it does not succeed in getting higher recall values, and there remains false negative cases. This is confirmed by Fig. 10, where the minimum missing rate for Faster R − CNN ∩ YOLO v2 detector is higher than other parallel integration methods.

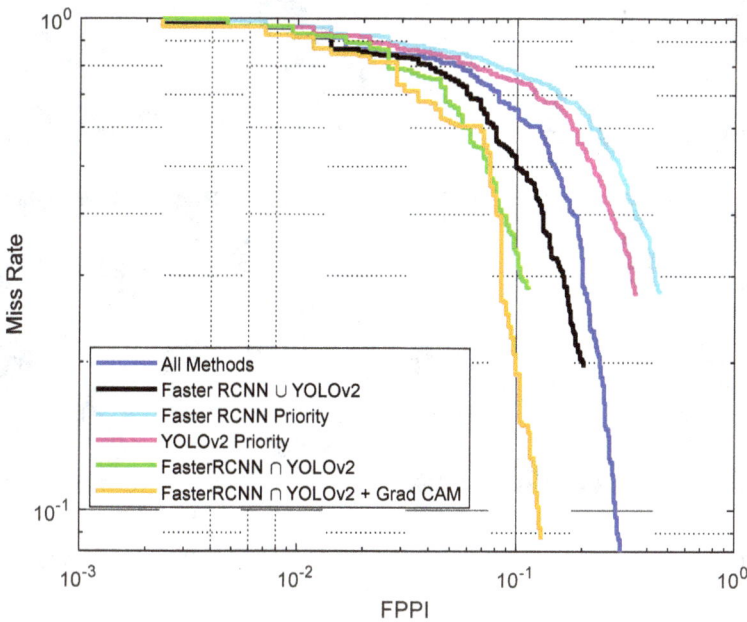

Fig. 10. Miss Rate Versus False Positive Per Image Curves for Different Parallel Integration Techniques.

Fig. 11 to Fig. 13 demonstrate examples of abnormality detection results through different methods. Fig. 11a and b illustrate detection by YOLO v2 and Faster R-CNN, where the ground truth bounding box is colored in blue while the detection outputs from the deep abnormality detectors are highlighted in green.

This example shows a case where both YOLO v2 and Faster R-CNN agree in detection (i.e. the detection bounding boxes by the two methods have an overlapping greater than 0.5) and thus the output of the linear regression model is considered as the final decision by parallel integration (all methods in Section 4.2) (Fig. 11c). Fig. 11d illustrates the result from Grad-CAM (th=0.9). One can notice that the detection from Grad-CAM is correctly pointing toward the lesion location with some margin of error.

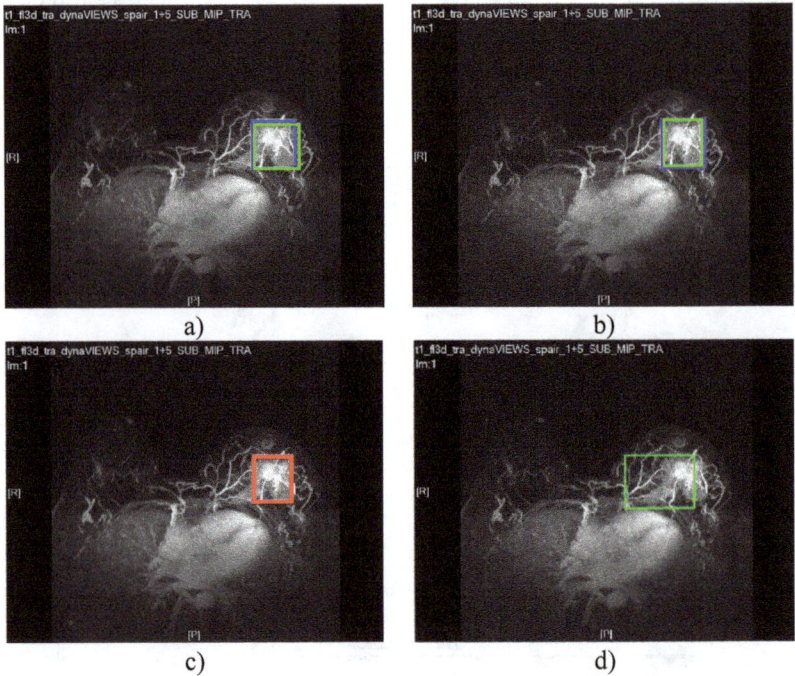

Fig. 11. Example of Success of Both YOLO v2 and Faster R-CNN (a) YOLO v2 (b) Faster R-CNN (c) Grad-CAM (d) Integrated Solution.

Fig. 12 illustrates an example where YOLO v2 fails to detect the abnormality (no green bounding box in Fig. 12a, the blue bounding box shows the ground-truth) while Faster R-CNN detects it (Fig. 12b). According to the proposed solution in Section 4.2, the detection by Faster R-CNN is considered as the final decision for parallel integration (Fig. 12c). Fig. 12d demonstrates the detection performed by Grad-CAM detector (th=0.9). It is then fed into a YOLO v2 for detecting the abnormality as in Fig. 12e.

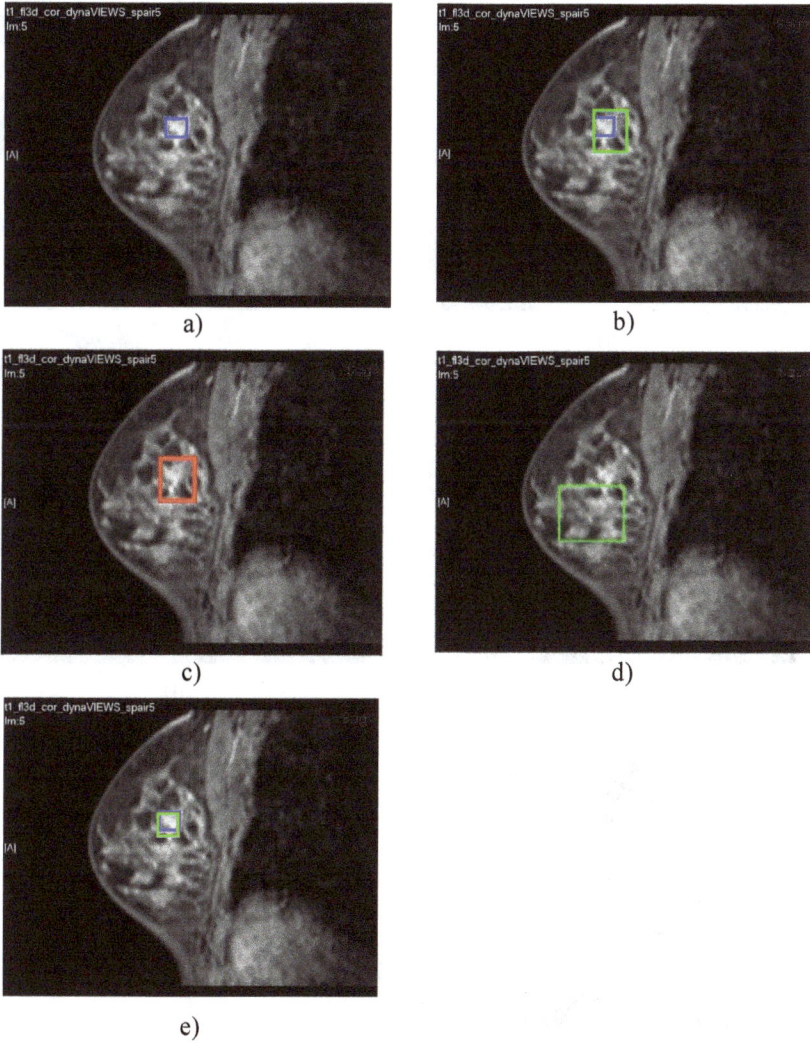

Fig. 12. An Example of Failure of YOLO v2: (a) YOLO v2 (b) Faster R-CNN (c) Parallel integration (all methods) (d) Grad-CAM (e) Cascaded integration (Grad-CAM+ YOLO v2).

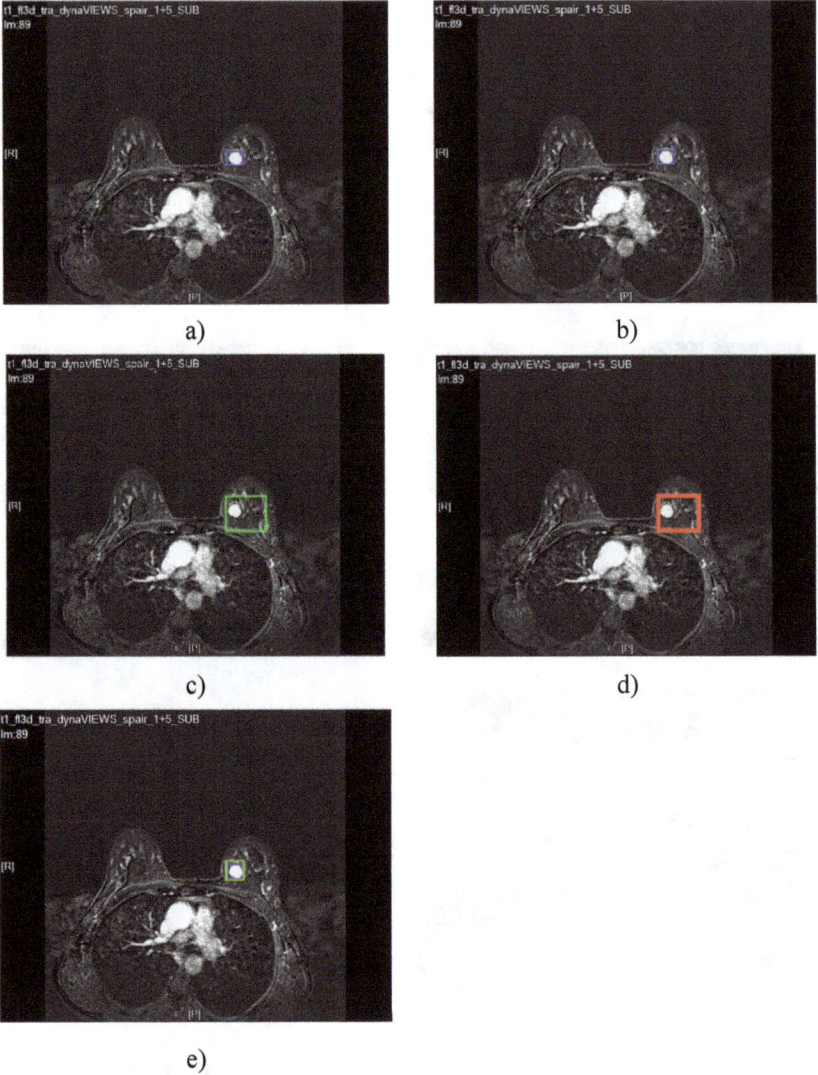

a)

b)

c)

d)

e)

Fig. 13. An Example of Failure of Both YOLO v2 and Faster R-CNN. (a) YOLO v2 (b) Faster R-CNN (c) Grad-CAM (d) All Method Parallel Integrated Solution, (e) Cascaded Integration Grad-CAM+YOLO v2.

Fig. 13 shows an example where neither YOLO v2 nor Faster R-CNN detects the lesion (Fig. 13a and b, the blue bounding box is the ground truth) while Grad-CAM succeeds to give a coarse detection. In the case of parallel integration (all methods as described in Section 4.2) when there is no detection from YOLO v2 or Faster R-CNN, the output of Grad-CAM (th=0.9) is chosen as the final detection (Fig. 13d). Fig. 13e depicts the detection result as performed by cascaded Grad-CAM+YOLO v2; however, the cascaded Grad-CAM + Faster R-CNN remains unsuccessful to detect the abnormality.

6   Conclusion

This chapter proposes different integration strategies using deep object detectors for the task of lesion detection in breast MRIs. It employs YOLO v2, Faster R-CNN and Grad-CAM, each with three different CNNs as backbone architecture, namely ResNet 50, GoogLeNet and MobileNet v2. Grad-CAM shows explainability of a deep model and is used to highlight the most important regions within an MRI for the classification task.

The performance of each abnormality detector is first studied on our dataset to determine the best performing CNN for each detector. Two integration approaches are proposed for boosting the performance. All approaches are evaluated on average precision and missed rate.

The first approach, referred to as cascaded integration, applies the Grad-CAM to a full MRI to make a coarse abnormality detection. A Faster R-CNN or YOLO v2 is successively applied to determine a fine bounding box around the lesion area. The cascaded integration slightly increases the average precision for both YOLO v2 and Faster R-CNN.

The second integration approach relies on final decision of all the three detectors and thus referred to as parallel integration. Six combination strategies are proposed for the parallel integration. For MRIs where a lesion is detected by both Faster R-CNN and YOLO v2, a linear regression model is trained to determine a more accurate bounding box for the lesion. To reduce the missed rate, all possible lesion areas detected by any of the studied approaches are merged to the final solution. Results confirm that a parallel integrated decision-making paradigm can both outperform each method in terms of average precision (Faster R-CNN ∩ YOLO v2 + Grad-CAM) and reduce the number of missed cases (All methods).

Based on the research findings in the chapter, a CADe system can be developed to detect high-risk lesions with more precise bounding boxes through matching detections by Faster R-CNN and YOLO v2. Other suspicious lesions can be determined by the overall integration approach for further analysis.

References

1. C. D. Lehman *et al.*, "National Performance Benchmarks for Modern Screening Digital Mammography," *Radiology*, vol. 283, no. 1, pp. 49–58, 2017, doi: 10.1148/radiol.2016161174.
2. P. Meyer, V. Noblet, C. Mazzara, and A. Lallement, "Survey on deep learning for radiotherapy," *Comput. Biol. Med.*, vol. 98, no. May, pp. 126–146, 2018, doi: 10.1016/j.compbiomed.2018.05.018.
3. A. Mehranian, H. Arabi, and H. Zaidi, "Vision 20/20: Magnetic resonance imaging-guided attenuation correction in PET/MRI: Challenges, solutions, and opportunities," *Med. Phys.*, vol. 43, no. 3, pp. 1130–1155, 2016, doi: 10.1118/1.4941014.
4. T. Zhou, S. Ruan, and S. Canu, "A review: Deep learning for medical image segmentation using multi-modality fusion," *Array*, vol. 3–4, no. May, p. 100004, 2019, doi: 10.1016/j.array.2019.100004.
5. L. Shen, L. R. Margolies, J. H. Rothstein, E. Fluder, R. McBride, and W. Sieh, "Deep Learning to Improve Breast Cancer Detection on Screening Mammography," *Sci. Rep.*, vol. 9, no. 1, pp. 1–12, 2019, doi: 10.1038/s41598-019-48995-4.
6. S. Han *et al.*, "A deep learning framework for supporting the classification of breast lesions in ultrasound images," *Phys. Med. Biol.*, vol. 62, no. 19, pp. 7714–7728, 2017, doi: 10.1088/1361-6560/aa82ec.
7. S. P. Rana *et al.*, "Machine Learning Approaches for Automated Lesion Detection in Microwave Breast Imaging Clinical Data," *Sci. Rep.*, vol. 9, no. 1, pp. 1–12, 2019, doi: 10.1038/s41598-019-46974-3.
8. R. Ha *et al.*, "Prior to Initiation of Chemotherapy, Can We Predict Breast Tumor Response? Deep Learning Convolutional Neural Networks Approach Using a Breast MRI Tumor Dataset," *J. Digit. Imaging*, vol. 32, no. 5, pp. 693–701, 2019, doi: 10.1007/s10278-018-0144-1.
9. R. Ha *et al.*, "Predicting Breast Cancer Molecular Subtype with MRI Dataset Utilizing Convolutional Neural Network Algorithm," *J. Digit. Imaging*, vol. 32, no. 2, pp. 276–282, 2019, doi: 10.1007/s10278-019-00179-2.
10. P. Herent *et al.*, "Detection and characterization of MRI breast lesions using deep learning," *Diagn. Interv. Imaging*, vol. 100, no. 4, pp. 219–225, 2019, doi: 10.1016/j.diii.2019.02.008.
11. K. He, X. Zhang, S. Ren, and J. Sun, "Deep Residual Learning for Image Recognition," in *2016 IEEE Conference on CVPR*, 2016, vol. 2016-Decem, pp. 770–778, doi: 10.1109/CVPR.2016.90.
12. G. Amit, R. Ben-Ari, O. Hadad, E. Monovich, N. Granot, and S. Hashoul, "Classification of breast MRI lesions using small-size training sets: comparison of deep learning approaches," *Med. Imaging 2017 Comput. Diagnosis*, vol. 10134, no. March 2017, p. 101341H, 2017, doi: 10.1117/12.2249981.
13. K. Simonyan and A. Zisserman, "Very deep convolutional networks for large-scale image recognition," *3rd Int. Conf. Learn. Represent. ICLR 2015 - Conf. Track Proc.*, pp. 1–14, 2015.
14. H. Whitney, H. Li, J. Yu;, P. Liu, and M. L. Giger, "Comparison of Breast MRI Tumor Classification Using Radiomics , Transfer Learning From Deep Convolutional Neural Networks, and Fusion Methods," pp. 1–15, 2019, doi: 10.1109/JPROC.2019.2950187.

15. M. A. Al-masni *et al.*, "Simultaneous detection and classification of breast masses in digital mammograms via a deep learning YOLO-based CAD system," *Comput. Methods Programs Biomed.*, vol. 157, pp. 85–94, 2018, doi: 10.1016/j.cmpb.2018.01.017.

16. J. Redmon, S. Divvala, R. Girshick, and A. Farhadi, "You only look once: Unified, real-time object detection," *Proc. IEEE Comput. Soc. Conf. Comput. Vis. Pattern Recognit.*, vol. 2016-Decem, pp. 779–788, 2016, doi: 10.1109/CVPR.2016.91.

17. J. Y. Chiao, K. Y. Chen, K. Y. K. Liao, P. H. Hsieh, G. Zhang, and T. C. Huang, "Detection and classification the breast tumors using mask R-CNN on sonograms," *Medicine (Baltimore).*, vol. 98, no. 19, p. e15200, 2019, doi: 10.1097/MD.0000000000015200.

18. K. He, G. Gkioxari, P. Dollar, and R. Girshick, "Mask R-CNN," *Proc. IEEE Int. Conf. Comput. Vis.*, vol. 2017-Octob, pp. 2980–2988, 2017, doi: 10.1109/ICCV.2017.322.

19. S. Ren, K. He, R. Girshick, and J. Sun, "Faster R-CNN: Towards Real-Time Object Detection with Region Proposal Networks," *IEEE Trans. Pattern Anal. Mach. Intell.*, vol. 39, no. 6, pp. 1137–1149, 2017, doi: 10.1109/TPAMI.2016.2577031.

20. M. Fan, Y. Li, S. Zheng, W. Peng, W. Tang, and L. Li, "Computer-aided detection of mass in digital breast tomosynthesis using a faster region-based convolutional neural network," *Methods*, vol. 166, no. December 2018, pp. 103–111, 2019, doi: 10.1016/j.ymeth.2019.02.010.

21. J. Redmon and A. Farhadi, "YOLO9000: Better, Faster, Stronger," in *2017 IEEE Conference on Computer Vision and Pattern Recognition (CVPR)*, 2017, vol. 2017-Janua, pp. 6517–6525, doi: 10.1109/CVPR.2017.690.

22. R. R. Selvaraju, M. Cogswell, A. Das, R. Vedantam, D. Parikh, and D. Batra, "Grad-CAM: Visual Explanations from Deep Networks via Gradient-Based Localization," in *2017 IEEE International Conference on Computer Vision (ICCV)*, 2017, vol. 2017-October 2017, pp. 618–626, doi: 10.1109/ICCV.2017.74.

23. R. Girshick, J. Donahue, T. Darrell, and J. Malik, "Rich feature hierarchies for accurate object detection and semantic segmentation," *Proc. IEEE Comput. Soc. Conf. Comput. Vis. Pattern Recognit.*, pp. 580–587, Nov. 2013, doi: 10.1109/CVPR.2014.81.

24. J. R. R. Uijlings, K. E. A. Van De Sande, T. Gevers, and A. W. M. Smeulders, "Selective search for object recognition," *Int. J. Comput. Vis.*, vol. 104, no. 2, pp. 154–171, 2013, doi: 10.1007/s11263-013-0620-5.

Chapter 2

Intensive Survey on Peripheral Blood Smear Analysis Using Deep Learning

Rabiah Alqudah and Ching Y. Suen

Dept. of Computer Science, Concordia University
r_alquda@ecs.concordia.ca, suen@ecs.concordia.ca

Abstract: Peripheral Blood Smear (PBS) analysis is a routine test carried out in specialized medical laboratories by specialists to assess some aspects of health status that are measured and assessed through blood. PBS analysis is prone to human errors and the usage of computer-based analysis can greatly enhance this process in terms of accuracy and cost. Despite the challenges, Deep Learning neural networks have shown impressive performance in this context. In this study the recent contributions are summarized along with the main challenges and future directions in this context.

Keywords: Automated blood smear analysis, Computer aided diagnosis, Deep learning, Leukemia diagnosis

1 Introduction

Blood cell analysis is a vital source of information for Medical doctors to diagnose patients for certain diseases. They provide important indicators of our health status. A blood test must be performed in a specialized Medical laboratory by specialists. There are three major types of blood cells: Red Blood Cells (RBCs), White Blood Cells (WBCs), and Platelets. RBCs, also known as erythrocytes, are the most common blood cell type and are responsible primarily for carrying oxygen and carbon dioxide to the entire body. WBCs, also known as leukocytes, are the primary defense system against infectious disease and foreign invaders. WBCs are much less common than RBCs. For example, the number of WBCs in adult males ranges from 4.5 to 11.5 thousand in 1 microliter, where the number of RBCs in adult males ranges from 4.6 to 6 million in 1 microlitre [1]. Platelets, also known as thrombocytes, are non-nucleated entities and are 2 to 4 micrometers in diameter, and is responsible for repairing blood vessels in case of injury.

A Peripheral Blood Smears (PBS), also known as a blood films, is the result of dispersing and staining a thin layer of blood on a microscope slide that is made of glass. Typically, it is hard, even for experts, to categorise some abnormal blood cells on the slide. PBSs are used to verify the results obtained by automated analyzer tools, to identify abnormal, immature, and/or atypical cells, and to recognize morphological abnormalities that are beyond the capabilities of automated analyzers. Blood smear analysis is a daily, time consuming procedure for lab specialists, hence, the automation of blood smear analysis has attracted the attention of researchers in recent years, and despite the good results achieved so far, many challenges still arise.

Image processing techniques ([2], [3], [4]), Machine Learning techniques ([5], [6], [7]), and Deep Learning techniques have been widely employed in analyzing blood samples and diagnosing abnormalities. In this work, a comprehensive survey of the recent work done in PBS analysis using deep learning networks is presented. The rest of this paper is organized as follows: Section 2 presents the recent contributions in the area of blood smear analysis using deep learning and highlights the main findings and conclusions of them. Section 3 concludes the paper, discusses and suggests open research areas for future investigations. Finally, in Section 4 this work is concluded.

2   Related Works on Peripheral Blood Smear Analysis

Blood smear analysis is an important diagnostic tool for medical doctors. Recent advancements in machine and deep learning paved the way for researchers to utilize these learning networks for blood smear analysis. On the other hand, such learning algorithms are data hungry, and data-driven. Hence, having a balanced, expressive dataset is a key factor in having reliable, well generalized models. In the next section, available blood smears and blood cell datasets are listed.

2.1 *Available Datasets*

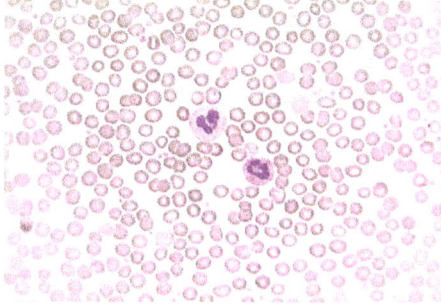

Fig. 1　Whole-slide images

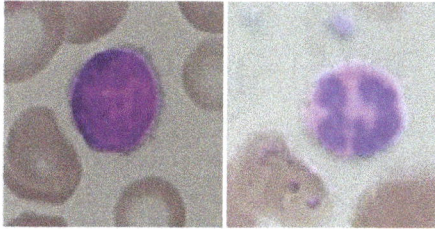

Fig. 2　Light microscopic images

In the context of PBS analysis research, two types of datasets can be found: light microscopic image datasets, a.k.a, blood cell datasets, and whole-slide datasets, as shown in Figures 2 and 1, respectively. The first type is being extensively investigated compared to the second one despite the fact that the whole-slide sets pose more realistic and challenging scenarios, as the entities in the blood appear microscopic, touching and crowded instead of the simple scenario represented in the light microscopic images where only one object of interest appears. Moreover, due to data scarcity and the strict security and privacy constraints enforced on blood smear datasets, some researchers opted to create synthetic blood smear datasets as a solution. Some public and synthetic datasets are listed in sections 2.1.1 and 2.1.2.

2.1.1 *Synthetic Blood Smear Datasets*

Synthetic datasets are widely used as a viable solution to data scarcity [8] [9]. Multiple techniques have been used to generate such datasets. As can be seen in Figure 1, the complexity of synthesizing blood smears mainly arises from the high number of blood cells that appear in each blood smear. Only a handful of works proposed methods to generate synthetic blood smears. The work in [10], generated blood smears by pasting blood cells on blood smear canvases by applying a Markov random process followed by a pix2pixHD network. A shortcoming of this synthetic dataset is that it only includes RBCs, which is a non-realistic scenario.

Another synthetic dataset was proposed in [11]. RBCs, WBCs, and Platelets were considered in this public dataset along with 15 morphological abnormalities. In this work the authors generated a balanced synthetic blood smear dataset by implementing an algorithm that utilises Locality Sensitive Hashing (LSH) as a backbone.

2.1.2 *Non-synthetic Blood Smear and Blood Cell Datasets*

Some public blood smear and blood cell datasets are:

(1) Acute Lymphoblastic Leukemia Image Database for Image Processing (ALL-IDB) dataset [12]: this dataset is composed of 108 images collected in September 2005. It contains approximately 39000 entities with an image resolution of 2592x1944. The dataset contains records for both healthy and sick people and it only supports annotation for blast cells. ALL-IDB1 dataset consists of whole-slide images, whereas ALL-IDB2 consists of light microscopic images, i.e., only one blood cell shows in each image. This dataset has been mainly utilised for Leukemia detection.

(2) Blood Cell Count and Detection (BCCD) dataset [13]: this dataset consists of 364 images. The dataset is annotated in VOC format for RBCs, WBCs, and Platelets. This dataset can be utilised for classifying and counting blood cells

(3) Leukocyte Images for Segmentation and Classification (LISC) dataset [14]: this dataset consists of 250 light microscopic images of the five main WBC types: Neutrophil, Lymphocyte, Monocyte, Eosinophil, and Basophil. The dataset also contains 126 whole-slide blood smears but without annotation. This dataset can be utilised for automating the differential blood test.

(4) National Institute of Health (NIH) dataset [15] [16]: this dataset was published in 2018. It consists of light microscopic cell images from thin blood smear slide in which images were collected and photographed at Chittagong Medical College Hospital, Bangladesh. The dataset contains 27,558 light microscopic cell images, with equal instances of Plasmodium parasitized and uninfected segmented red blood cell images. This dataset has been widely used in Malaria related research.

Some general conclusions can be drawn from this section are:

(1) A promising solution to the data imbalance and scarcity issues in the context of PBSs can generate more synthetic data. However, Only a handful of works have provided synthetic data generation solutions in this context.
(2) Most public datasets are composed of light microscopic images, or cropped blood smear images that only show small cropped portions of blood smears. Hence, more whole-slide blood smear datasets are needed as it poses more realistic and challenging scenarios.
(3) The majority of the public datasets are only annotated for the main blood cell types (RBCs, WBCs, Platelets) or only the main WBC or RBC subtypes. On the other hand, datasets annotated for morphological abnormalities are scarce. This has resulted in limiting most of the work done in this context to only consider the main blood cell types and subtypes. If annotations are made to cover more blood cell subtypes and morphological abnormalities, then computer researchers will be able to tackle more areas of blood analysis.

Computer researchers have mainly focused on three distinct directions in the context of blood smear analysis:

(1) Malaria Detection
(2) Blood Cell Detection and Classification
(3) Leukemia Diagnosis

In sections 2.2, 2.3, and 2.4, we present the recent state-of-the-art techniques used for each of these directions in greater details.

2.2 *Malaria Detection*

Malaria is a life-threatening disease caused by parasites that are transmitted to people through the bites of infected female Anopheles mosquitoes, The estimated number of malaria deaths stood at 435,000 in 2017. One of the most accurate methods to diagnose malaria is by examining a PBS to look for Plasmodium falciparum parasites infecting some RBCs. Many studies used the NIH dataset to train their neural networks ([17], [18], [19], [20], [15], [21]).

In [17], a Convolution Neural Network (CNN) architecture that is comprised of 12 layers was trained with the NIH dataset after applying a set of augmentation operations such as horizontal flip, vertical flip, width shift, height shift, fill mode, zoom range, and rotational range. The dataset was preprocessed by applying normalization, gamma correction, and logarithmic correction to improve brightness and adjust contrast operations. This model achieved an accuracy of 98.23% and F1 score of 97.74%. In [18], the NIH dataset instances were mean normalized, and several augmentation techniques including rotations, translation, shearing, zooming, and flipping were performed. A CNN that consists of 3 blocks was trained for Malaria detection. Moreover Visual Geometry Group-19 (VGG-19), SqueezeNet, InceptionResNet-V2 were customized by truncating them at their deepest convolutional layer and adding a Global Average Pooling (GAP) and dense layers. Several combinations of the listed models were ensembled by taking the average of the predictions, VGG-19 and SqueezeNet; combination outperformed the individual models and other ensembles in all performance metrics with an accuracy of 99.51%.

The work in [19] utilized the same NIH set to train a CNN of 3 convolutional layers, one hidden layer, input, flatten and output layers. This shallow CNN achieved a good accuracy of 95% with no augmentation or preprocessing. In [20], the NIH dataset was preprocessed by stain normalization, Min-Max Normalization, and Standardization. Many augmentation techniques have been applied: horizontal and vertical flips, Gaussian blur, rotation, horizontal and vertical shifting, darkening and lightening, ZCA whitening, and feature wise standardization, and Change of color space and Gaussian Blur. The dataset size was extended to 137,940 after augmentation. This work proposed a CNN that consists of 8 convolution layers. A VGG16 deep network was also customized by removing the pretrained fully convolution layers, and adding a dense layer, a dropout layer, and a fully connected layer. A third architecture called CNNEx-SVM was trained by emitting the customised VGG16 features to an Support Vector Machine (SVM). Finally, all models were ensembled by taking a weighted average of all predictions. The customised VGG16 achieve an accuracy of 97.6%, and the ensemble one achieved an accuracy of 97.7%.

In [15], a multi-scale Laplacian of Gaussian (LoG) filter was applied to whole-slide images from the NIH dataset to detect RBCs centroids. The detected cells were then segmented. Morphology opening operation is then applied to remove artifacts. F1-score of 0.952 was achieved in the detection phase. For cell classification a custom CNN of three convolutional layers and two fully connected layers and other pre-trained CNNs (AlexNet, VGG-16, Xception, DenseNet-121, and ResNet-50) were employed. ResNet outperformed the other networks with 95.7% accuracy. The accuracy was improved to 95.9% by evaluating the optimal layer for feature extraction, as the final layer isn't necessarily the optimal one.

The work in [21], an autoencoder was employed as a classifier. Multiple augmentation techniques were perfomed on the dataset, such as shift, zoom, and rotations. The performance of the autoencoder achieved an accuracy of 99.23%. Moreover, the model was tested to work on smartphones without the need for internet access. Some general conclusions can be drawn from [17], [18], [19], [20], [21], and [15] since it was all trained using the same dataset:

(1) Preprocessing and augmenting the dataset has helped in improving the testing results.
(2) Ensemble Learning when accompanied with preprocessing and augmentation can be a very effective and outperforms individual deep models.

Other works considered different datasets for training and testing ([22], [23], [24], [25], [26], [27], [28]). For example, the authors in [22] developed an Android smartphone application using a dataset of 1819 whole slide thick smear images from 150 patients. The parasite in this work is detected using a pipeline that starts by applying an intensity-based Iterative Global Minimum Screening (IGMS) procedure to reduce the size of the initial search space and limit the number of regions of interest which are fed to a CNN model consisting of seven convolutional layers. The classification accuracy was 93.46%.

The work in [23] annotated over 92k objects of the four major malaria species. Quality control is applied to each instance. Firstly, by checking the standard deviation of grayscale pixel values and the dynamic range of the gradient of the grayscale against preset thresholds. If either of these values is lower than the threshold it will be rejected. Secondly, focus metrics are calculated on the instance and on a corresponding artificially blurred version. Next, RBCs are detected and counted by applying binary grayscale clustering; candidate objects are chosen by finding connected-components in the threshold image. The pipeline then splits into two branches, one for rings (for quantitation), and one for late stages (for species ID). Each branch has two parts: First, a Gradient-boosted Tree (GBT) classifier was trained to act as a high-sensitivity distractor. Detected late stage parasites also pass through a species ID module. Finally, the various outputs are combined to deliver patient-level quantitation and species ID predictions. Two CNNs were proposed for classification, one for the ring branch and a second one for the late stage. The ring branch CNN had 3 convolutional layers, followed by two Inception modules and one fully connected layer where the late stage branch CNN was a fully convolutional neural network with 7 convolutional layers.

The work in [24] combined two datasets, a one collected by the authors and another public dataset from the Institute for Molecular Medicine Finland (FIMM). It has a collection of digital images infected by P.falciparum. The overall collection has a total of 1030 images of infected cells, and 1520 images of non-infected ones. The VGG network proposed in [24] was customized by removing the last three layers. The features obtained from the network are then fed to an SVM, and the model is called VGG19-SVM. Accuracy of 93.13% and F-score 91.66% were achieved. In [25], a dataset of 1000 instances, multi-wavelength was utilized to increase the sample size, 45°, and 135° rotations were also applied. The authors opted to utilize AlexNet, VGG-16, ResNet50, GoogLeNet, and a customized CNN network of 5 convolutional layers and 2 fully connected layers. ResNet outperformed with an accuracy of 97.6% in classifying the test set as healthy or infected.

A CNN framework is presented in [26] that is able to perform the extended depth of field images from z-stacks of thick blood films for automated malaria diagnosis. Two deep architectures were proposed, EDoF-CNN-3D, and EDoF-CNN-Max. In EDoF-CNN-3D, the encoder part of the network was modified by replacing the two-dimensional convolutions with three-dimensional ones. The output tensor was flattened on the z-axis before the residual layers of the network by average pooling. Where in EDoF-CNN-Max combines the idea behind the Siamese networks and the one behind the wavelet-based MSD EDoF. Each focal plane is passed through the encoder part of the network, and the maximum of the activation values are selected before going through the residual layers. The detection recall of the EDoF-CNN-3D method is 73%. While the work in [27] used other whole slides dataset which consists of 800 infected cells and 2000 healthy cells. Two augmentation techniques were applied on half of the dataset to produce two augmented sub-datasets: Image interpolation in the spatial domain (For any two images A and B in the dataset, we can generate a new image by finding a weighted average C), and image interpolation in the feature domain (For any two images A and B in the dataset, two 30-point feature vectors FA and FB can be obtained by the stacked autoencoders that have been trained we can generate a new 30-point vector FC by finding a weighted average).

The authors in [28] used 4100 whole slide peripheral blood smear images to train a Deep Belief Network (DBN) to classify the objects to either parasites or non-parasites. The objects are extracted from peripheral blood smear images using the level set method. A concatenated feature of color (histogram-based features and color coherence vector) and texture (Haralick features, LBP features, and gray level run length matrix feature) were used to initialize the visible layer of the 4 hidden layer DBN. The deep network achieved an F-score of 89.66.

The work in [29], trained the proposed framework on blood cell images dataset of size 1182. The proposed framework utilised a functional link artificial neural network (FLANN) and a sparse stacked autoencoder. The proposed model score an accuracy of 89.10%.

Although the models in studies [22] to [29] were trained with different datasets, however, a general finding is that using very deep neural networks, such as ResNet, can significantly improve the overall detection results.

2.3   *Blood Cell Detection and Classification*

The work in [30] proposed an architecture for microcytic hypochromia. The target features were a combination of blood smear image features extracted by AlexNet deep convolutional neural network and clinical features from (Red blood cell count (RBC), Haemoglobin concentration (HB), Red blood cell distribution width (RDW)). Samples were collected especially for this research from twenty patients. Both Principal Component Analysis (PCA), and Linear Discriminant Analysis (LDA) were used to reduce the feature set with minimal loss of information. k-Nearest Neighbors (k-NN), SVM, and Neural Network (NN), were employed for the classification phase. Each model was trained with three different feature sets: the clinical features, image features and fused features. The neural network and the SVM classifier scored 99% accuracy at testing when trained with the fused features, which shows the superiority of the proposed fusion model.

The work in ([31], [32]) classifies WBCs not only to their main types but also to some morphological abnormalities. In [31], a total of 14,700 annotated whole-slide images that include 11 categories of leukocytes were considered. Cells recognition was performed using Single Shot Detector (SSD) and YOLO3, Different variations of SSD and YOLOv3 were examined, 0.931 MaP and accuracy of 90.09% score were reported for the SSD 300 × 300, and the highest MaP scored with YOLOv3 320 × 320 is 0.92.

In [32], the authors collected a private dataset that contains a total of 92480 leukocytes belonging to 40 categories, with one object of interest in each instance. To handle the dataset imbalance, many augmentation techniques were applied such as, horizontal and vertical flips, and adding random noises and color changes to the original images. The architecture of the proposed deep residual neural network consists of 7 convolutional layers, 2 fully connected layers and three residual blocks to improve its performance. The authors examined 7 different schemes by using different activation functions to train the network. The average classification accuracy was 76.84%.

The work in [33] studied blood-cell classification in medical hyperspectral imaging (MHSI). It utilized four different architectures: SVM, typical VGG16, CNN without Gabor wavelet, CNN with Gabor wavelet and a combination of modulated Gabor wavelet and CNN kernels, named as MGCNN. In MGCNN, each convolutional layer performs a a dot product between multi-scale and orientation Gabor operators and the initial CNN kernels, to transform the convolutional kernels into the frequency domain in order to extract the features. Three datasets were utilized for testing, (1) Bloodcells1-3: CNN achieved the lowest overall accuracy (OA) of 81.35% and the highest OA is achieved using the proposed model with a score of 94.03%, (2) Bloodcells2-2 where 88.70%, and 94.40% were scored by CNN and the proposed model, respectively, (3) white blood cells dataset: SVM with PCA scored the lowest with overall accuracy of 90.83% and the proposed model achieved the highest accuracy score of 97.65%.

In [34], a subset of the All-IDB1 whole slide image dataset was used, as the authors selected 42 images and performed a pixel wise annotation on them. The training set size was increased from 29 to 145 images by performing random reflection and translation augmentation techniques. The class weighting technique was used to handle the dataset imbalance caused by having RBCs appear seven times more then WBCs. The set was then fed to a SegNet for semantic segmentation purposes. The highest accuracy was 94%, scored for the WBCs class.

In [35], a dataset available on GitHub of WBC images was used. Augmentation techniques such as random rotation, scaling, reflection, and shearing were performed. Gaussian noise was also applied to a subset of the training and testing set to train the network on poor quality images. The resulting 12,500 instances set is then fed to 3 sets of experiments that are made up of 10, 20, 30 CNNs where each CNN is constructed by generating random numbers of convolution blocks and layer sizes from preset ranges. The feature maps of each experiment were then concatenated and emitted to a PatternNet deep network to ensemble the strongest features to contribute in the final decision. The 30-CNNs experiment outperformed the 10 and 20 CNNs experiments with an accuracy score of 99.37%.

In [36] the input images are acquired from Pinterest online open source haematology database. RBCs are cropped from the blood smear to generate the dataset of normal, acanthocyte, sickle cell, teardrop and elliptocyte cell. The authors utilized SVM and AlexNet deep network. SVM model outperformed the AlexNet model, the authors referred this to the small dataset size. It is noted that there's a noticeable difference between the results of models, for example, the SVM model achieved 100% accuracy in classifying Achantocyte where the deep learning model achieved 0%.

The work in [37] combines Fourier Ptychographic Microscopy (FPM) and an improved version of You Only Look Once (YOLO) networks for WCB detection. In order to improve the detection of the microscopic WBCs, the feature maps of the last three layers are concatenated and passed to a final convolution layer. The proposed model was trained and tested on a 1000 whole slide image set. For the work in [38], the authors try to address the problem caused by the lack of some deep networks ability to fully exploit the long-term dependence relationship between certain key features of images and image labels. A combination of a CNN and Recurrent Neural Network (RNN) is employed to deepen the understanding of the image context. A dataset comprising 12,444 augmented and rotated images of blood cells were collected from Kaggle and BCCD public datasets. The proposed network consists of Pre-trained convolutional neural network layer, RNN layer, Merge layer, and fully connected layer with Softmax output. The proposed model achieved an accuracy of 90.79%.

In [39], a dataset of blood cell images was augmented by performing rotation, reflection and translation, three pipelines one for each blood cell were implemented to perform the classification using CNNs and UNet deep networks. The 64000 blood cells dataset used in [40] is a combination of all-idb, DPDx, ASH image bank and other images available on Google. The authors of [40] proposed a two-stage solution, In the first phase, a contour aware CNN was used for the segmentation of individual cells. In order to classify WBCs into five subtypes, features were extracted by a CNN and forwarded to ELM for classification. In order to classify RBCs into normal and identify abnormalities, several features were extracted such as centroid, medial axis ratio, and cell deform ratio, extracted features are forwarded to ELM for classification. Overall RBC classification accuracy was 90.10%, the highest WBC subtype accuracy was for Monocyte 98.68%.

The authors in [41] trained a capsule neural network on the LISC dataset to classify blood cells into the five main WBCs subtypes. In [42], not only a classification system was proposed, but also a different cell augmenting method was presented. The augmentation method was implemented by segmenting and pasting blood cells on different microscopic surrounding images. The model was able to achieve an accuracy score of 97.6%.

2.4 *Leukemia Diagnosis*

Leukemia is a fatal malignancy and has two main types: Acute and chronic which depend on how fast it progresses. Moreover, there are two subtypes of each leukemia main types depending on the size and the shape of the WBC: Lymphoid and myeloid. Acute leukemia is usually diagnosed after having clinical signs and symptoms that need to be confirmed by laboratory investigations. Complete blood count for WBCs, RBCs, platelets, and a peripheral blood smear are the initial tests. In many cases they will not be enough to confirm the diagnosis, which is why the clinical practice is to do a bone marrow smear and biopsy. A bone marrow specimen will have a smear and a biopsy. This specimen is usually good enough to confirm the diagnosis of leukemia, but more testing is mandatory for subtyping the leukemia into lymphoid or myeloid, and then subclassifying each subtype. In bone marrow specimen, number and shape of WBCs are the key point in diagnosing leukemia. Findings will be different between acute and chronic, myeloid and lymphoid, and each entity will have its own criteria of diagnosis.

The work in ([43], [44]) performs Leukemia diagnosis and classifies the result into its subtypes. In [43], two public leukemia datasets (ALL-IDB and ASH Image Bank) were used to train the network to classify the samples into one of the four main Leukemia types. The number of samples increased to 8 times for both datasets by applying shifting, rotation, and flipping. A CNN of 2 convolution layers, a Flatten layer, followed by a fully connected layer was proposed in this work. The accuracy of classification obtained was 81.74%.

The work in ([45], [46], [47], [48], [49]) aims to train a well generalized model to detect Leukemia. In [45] segmented white blood cell images of the

C-NMC dataset were augmented by performing horizontal and vertical flips, and random translations. A Squeeze-and-Excitation-ResNeXt50 network achieved a weighted F1-score of 88.91%. In [46] only one object of interest appears in each instance. all images were transferred to grayscale and the cell region was then binarized using the threshold estimated by Otsu's method followed by the erosion operation. The authors opted to train the model using a ResNet with two fully connected layers and utilize bagging ensemble training strategy. The model achieved an F1-score of 0.84.

The work in [47] trained three deep architectures (AlexNet, CaffeNet, Vgg-f) to generate features. The features space was then reduced by applying the gain ratio algorithm, before emitting the features to an SVM classifier. An accuracy score of 100% when the classification was performed by concatenating and reducing the features obtained by all models. In [48], CNN architecture comprising of 5 convolutional layers and 2 layers (fully connected and softmax) was trained on ALL-IDB1 dataset after applying many augmentation operations: histogram equalization, translation, reflection, rotation, shearing, conversion to grayscale, and blurring. The proposed method achieved an accuracy of 96.6%.

In [49], feature maps were extracted from the All-IDB1 dataset using AlexNet, CaffeNet, Vgg-f deep networks before being classified using SVM, Multilayer Perceptron (MLP) and Random Forest (RF). The authors also experimented the concatenation of feature maps obtained by all deep networks, the feature space was reduced by utilizing PCA technique and the majority voting rule to combine the outcomes obtained by each classifier. An accuracy score of 100% was achieved.

The authors in [50] utilised the ALL-IDB2 and the C-NMC datasets to train and test a novel framework for efficient feature selection. The proposed approach combined a VGG network with an improved version of the salp swarm algorithm (SSA). The VGG network was used as a feature extractor, where the SSA algorithm was employed for feature selection.

Three datasets, namely, ALL-IDB1, SMC-IDB, and IUMS-IDB were used for Leukemia detection in [51]. Each image instance was preprocessed by first applying an RGB to HSV conversion, followed by thresholding and Boolean mask generation. Next, a blob detection step based on the scale-normalized LoG (Laplacian of Gaussian) was executed. The resulting image was then segmented by applying multiple image processing filters and techniques. Next, An AlexNet network was employed for feature extraction. Finally the blood cells were classified using SVMs. The proposed framework scored an accuracy of 94.1%. A drawback of this framework, is the very long pipeline. A general conclusion can be highlighted from the literature in this subsection is that extracting features from an ensemble of deep networks and concatenating them before classification can be a very effective approach in detecting Leukemia. In Table 1 we summarize all the literature mentioned in this section.

Table 1: Summary of the reviewed literature

Reference	Method	Dataset	Results
Malaria Detection			
[21], (2020)	Autoencoder	Augmentation, NIH	Accuracy: 99.23%
[29], (2020)	FLANN and sparse stacked autoencoder	Private dataset	Accuracy: 89.10%.
[17], (2019)	CNN	Augmentation, NIH	Accuracy: 98.23%, F1 score: 97.74%.
[18], (2019)	CNN, VGG19, SqueezeNet, InceptionResNet-V2	Augmentation, NIH	Accuracy: 99.51%.
[19], (2019)	CNN	NIH	Accuracy: 95%
[20], (2019)	CNN, VGG16, SVM	Augmentation, NIH	Accuracy: 97.7%.
[15], (2018)	CNN, AlexNet, VGG-16, Xception, DenseNet-121, ResNet-50	NIH	Accuracy: 95.9%
[22], (2019)	CNN	Private dataset	Accuracy: 93.46%
[23], (2019)	CNN	Private dataset	Accuracy: ring: 94.8%,late: 96.6%

Table 1: Cont'd

[24], (2019)	VGG, SVM	Private dataset + FIMM	Accuracy: 93.13%, F-score 91.66%
[25], (2019)	AlexNet, VGG-16, ResNet50, GoogLeNet, CNN	Augmentation, Private dataset	Accuracy: 97.6%
[26], (2019)	CNN	Private dataset	Recall: 73%
[27], (2018)	CNN	Augmentation, Private dataset	Accuracy: 99%
[28], (2017)	Deep Belief Network (DBN)	Private dataset	F-score: 89.66%
Blood Cell Detection and Classification			
[41], (2020)	Caspule neural networks	LISC dataset	Accuracy: 96.86%
[30], (2019)	AlexNet, SVM	Private dataset	Accuracy: 99%
[31], (2019)	YOLO, SSD	Private dataset	Mean accuracy: 90.09%
[42], (2019)	CNN	Private dataset	Accuracy: 97.6%
[32], (2018)	Residual Network	Augmentation, Private dataset	Accuracy: 76.84%
[33] , (2019)	SVM, VGG16, CNN, CNN with Gabor wavelet	Private dataset	Accuracy: 97.65%
[34], (2017)	SegNet	Augmentation, ALL-IDB	Accuracy: 89.45%
[35], (2018)	CNN, PatterNet	Private dataset	Accuracy: 99.37%
[36], (2018)	SVM, AlexNet	Pinterest haematology dataset	Acanthocyte Accuracy: 100%
[37], (2018)	YOLO	Private dataset	Precision: 100%

Table 1: Cont'd

[38], (2018)	CNN, RNN	Augmentation, Kaggle and BCCD	Accuracy: 90.79%
[39], (2018)	CNN, U-Net	Augmentation, Private dataset	Specificity: 99.11%, Sensitivity: 100%
[40], (2017)	contour aware CNN, ELM	ALL-IDB, DPDx, ASH,Google.	RBC accuracy: 94.71%, WBC Accuracy: 98.68%
Leukemia Detection			
[51], (2020)	AlexNet	ALLIDB, SMC-IDB, IUMS-IDB	Accuracy: 94.1%
[45], (2019)	Squeeze-and-Excitation-ResNeXt50	Augmented, C-NMC dataset	F1-score: 88.91%
[43], (2019)	CNN	Augmentation, ALL-IDB and ASH Image Bank	Accuracy: 81.74%
[45], (2019)	Squeeze-and-Excitation-ResNeXt50	Augmented, C-NMC dataset	F1-score: 88.91%
[46], (2019)	ResNet	Private dataset	F1-score: 0.84
[47], (2018)	AlexNet, CaffeNet, VGG-f, SVM	Private dataset	Accuracy: 100 %
[48], (2018)	CNN	Augmentation, ALL-IDB1	Accuracy: 96.6%
[49], (2017)	AlexNet, CaffeNet, VGG-f, SVM	All-IDB1	Accuracy: 100%

3   Discussion and Future Directions

Despite the vast advancement in artificial intelligence and the robust frameworks presented in the PBS analysis context, many challenges remain to be tackled. Here, we list the main challenges that computer science researchers face when studying and developing blood analysis systems and some promising solutions:

(1) One of the biggest challenges is data scarcity. The number of datasets that are publicly available is limited, and many datasets are kept private which makes it harder to reproduce and improve the work presented in the literature. A possible viable solution for this challenge is creating more synthetic datasets that are not subject to privacy constraints.

(2) In the context of PBS analysis research, two types of datasets can be found:light microscopic images and whole-slide images. The latter dataset type is scarce, hence many PBS analysis application areas have not been completely explored using whole-slide images and there is still room for improvement in this context.

(3) Annotating PBS datasets is tedious and can only be done by medical experts. Moreover, it is still hard to find medical experts willing to devote time in the annotation process.

(4) The majority of the available datasets are only annotated for the main blood cell types, and a deeper level of annotation is still needed.

(5) PBS datasets are naturally imbalanced due to the natural distribution of WBCs and RBCs in blood. Imbalance could lead to a model that exhibits bias towards the majority class, and in some extreme cases would lead to ignoring the minority class altogether [52]. Imbalance in this context has been handled mostly by dynamic sampling approaches [53] and augmentation techniques such as rotation and translation. A drawback of these augmentation techniques is that it is performed by trial and error, and there is no formal method that can determine if an augmentation strategy will improve results until after training, which might lead to a time consuming training process [54].

It is also worth mentioning that many computer researchers still compare their work results to the manual methods of classifying and counting blood cells, where these manual methods are not in use anymore and are already replaced by certified automatic analyzers that can perform the task in seconds. Hence, counting and classifying normal blood components is not a current problem anymore, but counting and classifying abnormal cells is the problem that needs automated tools. If computers can read a slide and highlight number and shape and structure of all abnormal cells, this will make diagnosis and classification of leukemia for example an easy process in most cases.

Currently medical specialists need many stains and tools such as Flow-cytometry to determine the type, number, shape of cells which is time, money and effort consuming. Stains are very important diagnostic tools, they help in confirming and subtyping diseases in general. Staining even in laboratory automated devices have not reached the ideal status yet because of chances of procedure errors or human error. Laboratory services in diagnosis will step hugely forward if computers manage to do virtual staining accurately. This will save time, money and will reduce chances of procedure errors. We recognize that more work is still needed for the analysis of unstained slides as this direction is lacking investigation by computer researchers. The work in [55] is an example of a recent work in this direction.

4   Conclusion

A Peripheral Blood Smear (PBS), also known as a blood film, is the result of dispersing and staining a thin layer of blood on a microscope slide to assess some aspects of health status. Despite that several deep learning techniques and architectures have been proposed to automate this procedure, many areas are not tackled enough. In this study we present the recent contributions and discuss the promising future directions in the context of PBS analysis.

References

[1] B. F. Rodak, G. A. Fritsma and K. Doig, *Hematology: Clinical Principles and Applications.* Elsevier Health Sciences (2007), ISBN 9781416030065, https://books.google.com/books?id=6sfacydDNsUC.

[2] R. Hegde, K. Prasad, H. Hebbar and B. M. Singh, Image processing approach for detection of leukocytes in peripheral blood smears, *Journal of Medical Systems* **43**, 5 (2019), doi:10.1007/s10916-019-1219-3.

[3] S. Varma and S. Chavan, *Detection of Malaria Parasite Based on Thick and Thin Blood Smear Images Using Local Binary Pattern.* Springer, ISBN 978-981-13-1512-1, pp. 967–975 (2019), ISBN 978-981-13-1512-1, doi:10.1007/978-981-13-1513-8_98.

[4] R. Hegde, K. Prasad, H. Hebbar and B. M. Singh, Comparison of traditional image processing and deep learning approaches for classification of white blood cells in peripheral blood smear images, *Biocybernetics and Biomedical Engineering* **39**, 2, pp. 382–392 (2019), doi:10.1016/j.bbe.2019.01.005.

[5] M. M. Alam and M. T. Islam, Machine learning approach of automatic identification and counting of blood cells, *Healthcare Technology Letters* **6**, 4, pp. 103–108 (2019), doi:10.1049/htl.2018.5098.

[6] H. A. Aliyu, R. Sudirman, M. A. Abdul Razak and M. A. Abd Wahab, Red blood cell classification: Deep learning architecture versus support vector machine, in *2nd International Conference on BioSignal Analysis, Processing and Systems (ICBAPS)*. Kuching, Malaysia, pp. 142–147 (2018), doi:10.1109/ICBAPS.2018.8527398.

[7] H. A. Elsalamony, Detection of anaemia disease in human red blood cells using cell signature, neural networks and svm, *Multimedia Tools Appl.* **77**, 12, pp. 15047–15074 (2018), doi:10.1007/s11042-017-5088-9.

[8] M. Frid-Adar, I. Diamant, E. Klang, M. Amitai, J. Goldberger and H. Greenspan, Gan-based synthetic medical image augmentation for increased cnn performance in liver lesion classification, *Neurocomputing* (2018), doi:10.1016/j.neucom.2018.09.013.

[9] C. Han, H. Hayashi, L. Rundo, R. Araki, W. Shimoda, S. Muramatsu, Y. Furukawa, G. Mauri and H. Nakayama, Gan-based synthetic brain mr image generation, in *2018 IEEE 15th International Symposium on Biomedical Imaging (ISBI 2018)*, pp. 734–738 (2018).

[10] O. Bailo, D. Ham and Y. Min Shin, Red blood cell image generation for data augmentation using conditional generative adversarial networks, in *The IEEE Conference on Computer Vision and Pattern Recognition (CVPR) Workshops*. Long Beach, CA, USA (2019).

[11] R. Al-Qudah and C. Y. Suen, Synthetic blood smears generation using locality sensitive hashing and deep neural networks, *IEEE Access* **8**, pp. 102530–102539 (2020).

[12] R. Labati, V. Piuri and F. Scotti, All-idb: the acute lymphoblastic leukemia image database for image processing, in *18th IEEE International Conference on Image Processing*. Brussels, Belgium, pp. 2045–2048 (2011), doi:10.1109/ICIP.2011.6115881.

[13] S. G. account, Bccd dataset, https://github.com/Shenggan/BCCD_Dataset (2019).

[14] H. Rezatofighi and H. Soltanian-Zadeh, Automatic recognition of five types of white blood cells in peripheral blood, *Computerized medical imaging and graphics : the official journal of the Computerized Medical Imaging Society* **35**, pp. 33–43 (2011), doi:10.1016/j.compmedimag.2011.01.003.

[15] S. Rajaraman, S. Antani, M. Poostchi, K. Silamut, R. Maude, S. Jaeger and G. Thoma, Pre-trained convolutional neural networks as feature extractors toward improved malaria parasite detection in thin blood smear images, *PeerJ* **6**, e4568 (2018), doi:doi.org/10.7717/peerj.4568.

[16] NIH, National institute of health malaria dataset, https://lhncbc.nlm.nih.gov/publication/pub9932 (2018).

[17] R. Kumar, S. K. Singh and A. Khamparia, Malaria detection using custom convolutional neural network model on blood smear slide images, in *Advanced Informatics for Computing Research*. Singapore, pp. 20–28 (2019), doi:10.1007/978-981-15-0108-1-3.

[18] S. Rajaraman, S. Jaeger and S. K. Antani, Performance evaluation of deep neural ensembles toward malaria parasite detection in thin-blood smear images, *PeerJ* **7**, e6977 (2019), doi:10.7717/peerj.6977.

[19] S. C. Kalkan and O. K. Sahingoz, Deep learning based classification of malaria from slide images, in *2019 Scientific Meeting on Electrical-Electronics Biomedical Engineering and Computer Science (EBBT)*. Istanbul, Turkey, pp. 1–4 (2019), doi:10.1109/EBBT.2019.8741702.

[20] A. Rahman, H. Zunair, M. S. Rahman, J. Q. Yuki, S. Biswas, M. A. Alam, N. B. Alam and M. R. C. Mahdy, Improving malaria parasite detection from red blood cell using deep convolutional neural networks, (2019), arXiv:1907.10418.

[21] F. KMF, T. JF, S. MRA, M. S, M. N and R. T., Deep learning based automatic malaria parasite detection from blood smear and its smartphone based application, *Diagnostics* **10**, p. 329 (2020), doi:https://doi.org/10.3390/diagnostics10050329.

[22] F. Yang, M. Poostchi, H. Yu, Z. Zhou, K. Silamut, J. Yu, R. Maude, S. Jaeger and S. Antani, Deep learning for smartphone-based malaria parasite detection in thick blood smears, *IEEE Journal of Biomedical and Health Informatics* **PP**, 99, p. 1 (2019), doi:10.1109/JBHI.2019.2939121.

[23] C. Delahunt, R. Gebrehiwot, B. Wilson, E. Long, S. Proux, D. Gamboa, P. Chiodini, J. Carter, M. Dhorda, D. Isaboke, B. Ogutu, M. Jaiswal, W. Oyibo, E. Villasis, K. M. Tun, C. Bachman, D. Bell, C. Mehanian, M. Horning and G. Yun, Fully-automated patient-level malaria assessment on field-prepared thin blood film microscopy images, pp. 1–8 (2019), doi:10.1109/GHTC46095.2019.9033083.

[24] V. Arunagiri and R. B, Deep learning approach to detect malaria from microscopic images, *Multimedia Tools and Applications* **79**, pp. 15297–15317 (2019), doi:https://doi.org/10.1007/s11042-019-7162-y.

[25] N. Singla and V. Srivastava, Deep learning enabled multi-wavelength spatial coherence microscope for the classification of malaria-infected stages with limited labelled data size, *Optics & Laser Technology* **130**, p. 106335 (2020), doi:https://doi.org/10.1016/j.optlastec.2020.106335.

[26] P. Manescu, L. Neary-Zajiczek, M. J. Shaw, M. Elmi, R. Claveau, V. Pawar, J. Shawe-Taylor, I. Kokkinos, M. A. Srinivasan, I. Lagunju, O. Sodeinde, B. J. Brown and D. Fernandez-Reyes, Deep learning enhanced extended depth-of-field for thick blood-film malaria high-throughput microscopy, (2019), arXiv:1906.07496 [eess.IV].

[27] W. Pan, Y. Dong and D. Wu, *Classification of Malaria-Infected Cells Using Deep Convolutional Neural Networks*, chap. 8. IntechOpen, ISBN 978-1-78923-752-8, pp. 159–172 (2018), ISBN 978-1-78923-752-8, doi:10.5772/intechopen.72426.

[28] D. Bibin, M. S. Nair and P. Punitha, Malaria parasite detection from peripheral blood smear images using deep belief networks, *IEEE Access* **5**, pp. 9099–9108 (2017), doi:10.1109/ACCESS.2017.2705642.

[29] P. A. Pattanaik, M. Mittal and M. Z. Khan, Unsupervised deep learning cad scheme for the detection of malaria in blood smear microscopic images, *IEEE Access* **8**, pp. 94936–94946 (2020).

[30] S. Purwar, R. K. Tripathi, R. Ranjan and R. Saxena, Detection of microcytic hypochromia using cbc and blood film features extracted from convolution

neural network by different classifiers, *Multimedia Tools and Applications* (2019), doi:10.1007/s11042-019-07927-0.

[31] Q. Wang, S. Bi, M. Sun, Y. Wang, D. Wang and S. Yang, Deep learning approach to peripheral leukocyte recognition, *PloS one* **14**, 6 (2019), doi: 10.1371/journal.pone.0218808.

[32] F. Qin, N. Gao, Y. Peng, Z. Wu, S. Shen and A. Grudtsin, Fine-grained leukocyte classification with deep residual learning for microscopic images, *Computer Methods and Programs in Biomedicine* **162**, 8, p. 243–252 (2018), doi:10.1016/j.cmpb.2018.05.024.

[33] Q. Huang, W. Li, B. Zhang, Q. Li, R. Tao and N. H. Lovell, Blood cell classification based on hyperspectral imaging with modulated gabor and cnn, *IEEE Journal of Biomedical and Health Informatics* **24**, 1, pp. 160–170 (2019), doi:10.1109/JBHI.2019.2905623.

[34] T. Thanh, O.-H. Kwon, K.-R. Kwon, S.-H. Lee and K.-W. Kang, Blood cell images segmentation using deep learning semantic segmentation, in *2018 IEEE International Conference on Electronics and Communication Engineering (ICECE)*. Xian, China, pp. 13–16 (2018), doi:10.1109/ICECOME. 2018.8644754.

[35] J. L. Wang, A. Y. Li, M. Huang, A. K. Ibrahim, H. Zhuang and A. M. Ali, Classification of white blood cells with patternnet-fused ensemble of convolutional neural networks (pecnn), in *2018 IEEE International Symposium on Signal Processing and Information Technology (ISSPIT)*. Louisville, KY, USA, pp. 325–330 (2018), doi:10.1109/ISSPIT.2018.8642630.

[36] H. A. Aliyu, R. Sudirman, M. A. Abdul Razak and M. A. Abd Wahab, Red blood cell classification: Deep learning architecture versus support vector machine, in *2018 2nd International Conference on BioSignal Analysis, Processing and Systems (ICBAPS)*. Kuching, Malaysia, pp. 142–147 (2018), doi:10.1109/ICBAPS.2018.8527398.

[37] X. Wang, T. Xu, J. Zhang, S. Chen and Y. Zhang, So-yolo based wbc detection with fourier ptychographic microscopy, *IEEE Access* **6**, pp. 51566–51576 (2018), doi:10.1109/ACCESS.2018.2865541.

[38] G. Liang, H. Hong, W. Xie and L. Zheng, Combining convolutional neural network with recursive neural network for blood cell image classification, *IEEE Access* **6**, pp. 36188–36197 (2018), doi:10.1109/ACCESS.2018. 2846685.

[39] D. Mundhra, B. Cheluvaraju, J. Rampure and T. Rai Dastidar, Analyzing microscopic images of peripheral blood smear using deep learning, in *Deep Learning in Medical Image Analysis and Multimodal Learning for Clinical Decision Support*. Springer International Publishing, Cham, ISBN 978-3-319-67558-9, pp. 178–185 (2017), ISBN 978-3-319-67558-9.

[40] M. I. Razzak and S. Naz, Microscopic blood smear segmentation and classification using deep contour aware cnn and extreme machine learning, in *2017 IEEE Conference on Computer Vision and Pattern Recognition Workshops (CVPRW)*, pp. 801–807 (2017), doi:10.1109/CVPRW.2017.111.

[41] Y. Baydilli and U. Atila, Classification of white blood cells using capsule networks, *Computerized Medical Imaging and Graphics* **80**, p. 101699 (2020),

doi:10.1016/j.compmedimag.2020.101699.

[42] Y. Wang and Y. Cao, Human peripheral blood leukocyte classification method based on convolutional neural network and data augmentation, *Medical Physics* **47** (2019), doi:10.1002/mp.13904.

[43] N. A. Ahmed, A. Yiğit, Z. Isik and A. Alpkocak, Identification of leukemia subtypes from microscopic images using convolutional neural network, *Diagnostics* **9**, 3, p. 104 (2019), doi:10.3390/diagnostics9030104.

[44] S. Shafique and S. Tehsin, Acute lymphoblastic leukemia detection and classification of its subtypes using pretrained deep convolutional neural networks, *Technology in Cancer Research & Treatment* **17**, p. 1–7 (2018), doi: 10.1177/1533033818802789.

[45] J. Prellberg and O. Kramer, Acute lymphoblastic leukemia classification from microscopic images using convolutional neural networks, in A. Gupta and R. Gupta (eds.), *ISBI 2019 C-NMC Challenge: Classification in Cancer Cell Imaging*. Springer Singapore, Singapore, ISBN 978-981-15-0798-4, pp. 53–61 (2019), ISBN 978-981-15-0798-4.

[46] Y. Liu and F. Long, Acute lymphoblastic leukemia cells image analysis with deep bagging ensemble learning, *bioRxiv* (2019), doi:10.1101/580852.

[47] L. H. Vogado, R. M. Veras, F. H. Araujo, R. R. Silva and K. R. Aires, Leukemia diagnosis in blood slides using transfer learning in cnns and svm for classification, *Eng. Appl. Artif. Intell.* **72**, C, pp. 415–422 (2018), doi: 10.1016/j.engappai.2018.04.024.

[48] T. Thanh, C. Vununu, S. Atoev, S.-H. Lee and K.-R. Kwon, Leukemia blood cell image classification using convolutional neural network, *International journal of computer theory and engineering* **10**, 2, pp. 54–58 (2018), doi: 10.7763/IJCTE.2018.V10.1198.

[49] L. H. S. Vogado, R. D. M. S. Veras, A. R. Andrade, F. H. D. D. Araujo, R. R. V. e. Silva and K. R. T. Aires, Diagnosing leukemia in blood smear images using an ensemble of classifiers and pre-trained convolutional neural networks, in *2017 30th SIBGRAPI Conference on Graphics, Patterns and Images (SIBGRAPI)*. Campo Grande, Brazil, pp. 367–373 (2017), doi:10. 1109/SIBGRAPI.2017.55.

[50] A. Talaat, P. Kollmannsberger and A. Ewees, Efficient classification of white blood cell leukemia with improved swarm optimization of deep features, *Scientific Reports* **10** (2020), doi:10.1038/s41598-020-59215-9.

[51] C. Di Ruberto, A. Loddo and G. Puglisi, Blob detection and deep learning for leukemic blood image analysis, *Applied Sciences* **10**, p. 1176 (2020), doi: 10.3390/app10031176.

[52] J. M. Johnson and T. M. Khoshgoftaar, Survey on deep learning with class imbalance, *Journal of Big Data* **6**, 1, p. 27 (2019), doi:10.1186/ s40537-019-0192-5.

[53] M. Lin, K. Tang and X. Yao, Dynamic sampling approach to training neural networks for multiclass imbalance classification, *IEEE Transactions on Neural Networks and Learning Systems* **24**, 4, pp. 647–660 (2013), doi: 10.1109/TNNLS.2012.2228231.

[54] J. Lemley, S. Bazrafkan and P. Corcoran, Smart augmentation learning an

optimal data augmentation strategy, *IEEE Access* **5**, pp. 5858–5869 (2017), doi:10.1109/ACCESS.2017.2696121.

[55] T. Go, J. H. Kim, H. Byeon and S. J. Lee, Machine learning-based in-line holographic sensing of unstained malaria-infected red blood cells, *Journal of Biophotonics* **11**, 9, p. e201800101 (2018), doi:10.1002/jbio.201800101.

https://doi.org/10.1142/9789811239014_0003

Chapter 3

End-to-End Generative Adversarial Network for Hand-Vein Recognition

Huafeng Qins[1] and Mounim A. El-Yacoubi[2]

[1]Chongqing Engineering Laboratory of Detection Control and Integrated System, Chongqing Technology and Business University, Chongqing, China
qinhuafengfeng@163.com

[2]Telecom SudParis, Institut Polytechnique de Paris, France
mounim.el_yacoubi@telecom-sudparis.eu

Abstract: Although it has received increasing researchers' attention in recent years, palm-vein recognition still faces various challenges in practical applications, one of which is the lack of robustness against image quality degradation, resulting in reduction of the verification accuracy. To address this problem, this chapter proposes an end-to-end generative adversarial network to automatically extract the vein pattern network, thus without resorting to any hand-crafted segmentation techniques of grayscale images into vein pixels and background. Firstly, we label the palm-vein pixels based on a combination of handcrafted segmentation methods and reconstruct a training set accordingly. Secondly, an end-to-end vein segmentation model is proposed based on a generative adversarial network. After training, this model outputs an image map where each value is the probability that the corresponding pixel belongs to a vein pattern. The resulting map is then subject to binarization by thresholding and stored in a binary image, used subsequently for verification matching. The experimental results on the public CASIA and PolyU palm-vein datasets demonstrate the effectiveness of our proposed method.

Keywords: Palm-vein recognition, Generative adversarial network, U-Net, Convolutional Neural Network

1 Introduction

With the rapid development of digital economy and internet technology, traditional verification techniques based on passwords or smart cards can hardly meet the requirements of convenience, reliability, and security in practical applications. In this context, automatic biometric personal verification using physiological and/or behavioral characteristics has received increasing attention and has become one of the most critical and challenging tasks in information security. Currently, the modalities employed for personal verification are divided into two categories: (1) Extrinsic modalities: fingerprint [1], iris [2], gait [3], face [4], hand shape [5] and signature [6], and (2) Intrinsic modalities such as finger-vein [7], palm-vein [8], dorsal hand-vein [9]. The extrinsic biometric traits are prone to attack, causing some concerns on privacy and security in practice applications. For example, face and fingerprint are easy to acquire and their fake versions have been successfully employed to fool the recognition systems. Similarly, iris recognition systems are also prone to be attacked by the iris fake versions [10], [11].

Unlike the extrinsic biometric features, vein traits are concealed beneath human skin and thus are not easy to copy, which results in high security and privacy. As they are hidden, the veins are difficult to observe in visible light. Nonetheless, they can be captured by infrared illumination with wavelength of 850 nm [7], [8], [9]. Some medical research works [12,13] have shown that the structure of blood vessels has high uniqueness for each individual [13], and can distinguish even identical twins [12], [14]. Therefore, the vein authentication technology has been widely investigated in the past years [7], [8], [9], [15-27].

Compared with the finger-vein, the palm-vein has more complex lines and structures, that are beneficial to improve the verification rate. Therefore, the palm-vein verification technology has received more and more attention from researchers and industry.

Palm-vein recognition is still a challenging task because the acquisition process is inherently affected by many factors such as temperature, equipment, user habits, illumination, etc. As a result, the captured image includes noise and irregular shadow, which will ultimately degrade the performance of verification systems. To solve these problems, various methods are proposed to extract the palm-vein texture. The latter can be broadly classified into two categories:

(1) Handcraft-based segmentation approaches: Some researchers assume that the cross-profile of vein patterns shows a valley shape and build, accordingly, a mathematical model to detect valleys for vein segmentation. The representative approaches include line tracking methods [15], [16], and curvature-based measures [17], [18], [19], [20]. Other studies observe that the vein patterns show a line-like texture in a predefined neighborhood region, so Gabor filters [7], [21], matched filters [22], wide line detector [23] and neural networks [24], are proposed to extract vein textures.

(2) Deep learning-based segmentation methods: Convolutional neural networks (CNNs) have shown to outperform the state of the art in the computer vision field. Different from handcrafted methods, deep learning-based segmentation methods consist of an end-to-end architecture that does not require any a priori assumption on the distribution of the vein attributes. Some researchers have applied CNNs for vein segmentation [25-27] and vein quality assessment [28-30]. For example, Qin et al. [25] proposed

a CNN to segment vein patterns for verification. To improve the performance, an iterative deep neural network [26] is proposed for hand-vein verification. To address the wrong label problem, a generative adversarial network (GAN) [27] is employed to extract the finger-vein texture.

The handcrafted methods [7, 8, 9, 15-24] described above depend on the assumption distributions such as valleys and line segments. However, these assumptions may not be always effective because the vein pixel values may generate different distributions. Therefore, these methods do not always perform well in practical applications. The deep learning based methods [25-27] are capable of extracting the vein texture without requiring any assumption and have shown better performance than handcrafted approaches. However, they still have two critical issues. First, they divide an image into various patches and build a patch-based dataset to train a deep neural network for feature extraction. Therefore, they do not account for global correlations when processing individual patches, which may lead to failures caused by noise and local irregular shadow regions. In fact, there is a tradeoff between localization accuracy and the use of global context. Large patches require more max-pooling layers, which may degrade the localization accuracy, while small patches prevent the network from capturing the global context. Second, the classification strategy for each patch is computationally intensive for both the training and testing phases.

Generative Adversarial Networks (GANs) are a powerful class of generative models that cast generative modeling as a game between two networks: a generator network produces synthetic data given some noise source and a discriminator network discriminates between the generator's output and true data. The adversarial learning framework allows us to model the underlying distribution of plausible samples only from training data, without manually interacting with parameters controlling complex mathematical models. This framework was successfully applied for various computer vision tasks. Some researchers brought it into medical image segmentation field [31], [32] and harnessed it for retina image segmentation, brain segmentation, and neuronal membranes segmentation.

Inspired by this idea, we propose, in this chapter, an end-to-end vein image segmentation model for hand-vein verification. Our main contributions can be summarized as follows: 1) this work makes the first attempt to accommodate a generative adversarial network (GAN) on the hand-vein verification task. First, we propose a GAN-based model to segment the vein network for verification and design an appropriate loss function for this purpose. Our GAN consists of a discriminator and a generator. The discriminator is designed to distinguish a real vein pattern from the output of the generator. The generator is a U-net network that takes a whole vein image as its input and outputs the associated vein probability map. To generate more consistent images, the L_2 loss instead of L_1 loss is employed as loss function to penalize the distance between the ground truth and output. Secondly, unlike exiting vein segmentation approaches [25-27] that divide an image into several patches and construct a patch-based set for training, the entire hand-vein images and theirs corresponding ground truths (binary images) are, in our scheme, directly input into GAN for training. To generate labeled images for model training, we employ an existing handcrafted image segmentation approach to extract the vein network from an image and the resulting binary image is used to automatically label each pixel as vein/non-vein. After training, the entire image is taken as input of GAN to predict the probability of each of its pixels to belong to a vein pattern. Therefore, our approach takes into account vein localization and the use of global context at the same time. Moreover, it can directly predict binary vein patterns from vein images in one forward propagation without pre-processing and post-processing, which results in low-time cost. To perform training of the hand-vein generation module and the hand image to vein network mapping, the adversarial loss and the binary cross entropy loss are combined in our GAN model. The U-net allows to generate realistic vein networks by minimizing an adversarial loss, while the cross entropy loss guarantees that the output of U-net is globally consistent. Therefore, the proposed framework provides an effective end-to-end hand vein image segmented tool, capable of extracting vein networks from an input hand vein image. 2) We have carried out rigorous experiments to investigate the capacity of our vein segmentation model. Our experimental results show that the proposed model is capable of extracting the vein patterns from raw hand vein images and achieve better performance compared to existing approaches.

2   The Proposed Method

We propose a generative adversarial network to extract palm vein patterns from the grayscale images. To get the ground truth, given an image, various handcrafted methods are employed to segment the vein network and the resulting binary images are combined to label the vein pixels. The adversarial learning framework allows us to model the underlying distribution of plausible images only from training data, without manually interacting with parameters controlling complex mathematical models. Concretely, we build and train a generative adversarial network (GAN) to segment the vein patterns from the grayscale image. Our GAN consists of two networks, discriminator and generator. The discriminator is a CNN with four convolutional layers and one fully connection layer, and the U-Net is employed as generator. The discriminator attempts to find the features that distinguish fake binary vein images from real ones, while the generator learns to synthesize, from grayscale images, binary ones that are difficult for the discriminator to judge as real or fake. Given a grayscale image, a binary vein image is either produced through the baseline segmentation techniques or is synthesized through the generator. Combined with the grayscale image, the binary image produces a real pair in the first case and a fake pair in the second, as shown in Fig. 1. The discriminator takes a fake pair or a real pair as an input to determine whether the vein image is the gold standard from the three baselines or the output of the generator.

After adversarial training, given an image, either from a training or a test image, the generator computes the probability of each pixel to belong to a vein pattern and outputs

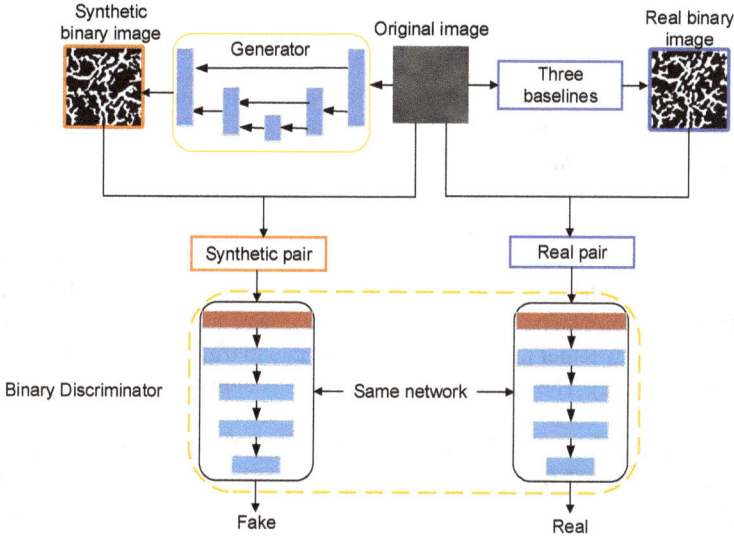

Fig. 1. Overview of our approach.

an enhanced image accordingly. Then, we encode the resulting enhanced image according to the winning class (based on a probability threshold of 0.5). Finally, we match the encoded vein patterns for verification.

2.1   Labeling vein patterns

Similar to work [25], we employ the Repeated line tracking [15], Gabor filters [7], and Hessian phase [9] techniques to segment the vein patterns in a given image I. We get in this way three corresponding binary vein images L_1, L_2, and L_3, where 1 and 0 denote the vein pixel and background pixel respectively. Finally, the three binary images are combined to obtain a labeled image L as follows.

$$L(i,j) = \begin{cases} 1 & \sum_{k=1}^{K} l_k(i,j) \geq \frac{K}{2} \\ 0 & \sum_{k=1}^{K} l_k(i,j) < \frac{K}{2} \end{cases} \qquad (1)$$

where K is the number of the segmentation approaches (three here). We use the labeled map $L(i,j)$. (0 and 1 denote background and vein pixels respectively) as the ground truth of the corresponding grayscale palm-vein image I and construct accordingly the training set.

3   GAN Framework Structure

GAN [33] is a specific framework of a generative model aiming at implicitly learning the data distribution p_{data} from a set of samples $(x_1, x_2, \ldots, x_m)$ (e.g. images) to further generate new samples drawn from the learned distribution. In our work, we have explored GAN for hand-vein recognition. The GAN consists of two modules i.e. generator and discriminator. The generator takes a grayscale hand image as input and generates a vein probability map with the same size. The values in the probability map, ranging from 0 to 1, indicate the probability of each pixel to belong to a vein pattern. The discriminator takes as input a pair of a grayscale hand image and the associated segmented vein image to learn whether the vein image is the ground truth (real) or rather the output of the generator (fake). The framework of our GAN is depicted in Fig. 1.

Generator Architecture. For the generator, we follow the spirit of U-Net [34] where the initial convolutional feature maps are skip-connected to upsampled layers from bottleneck layers. This skip-connection is crucial to segmentation tasks as the initial feature maps maintain low-level features such as edges and blobs that can be properly exploited for accurate segmentation. As shown in Fig. 2, the U-Net architecture consists of a contracting path and an expansive path. In the contracting path, down-sampling is a classical convolutional neural network, which consists of five layers. In each layer,

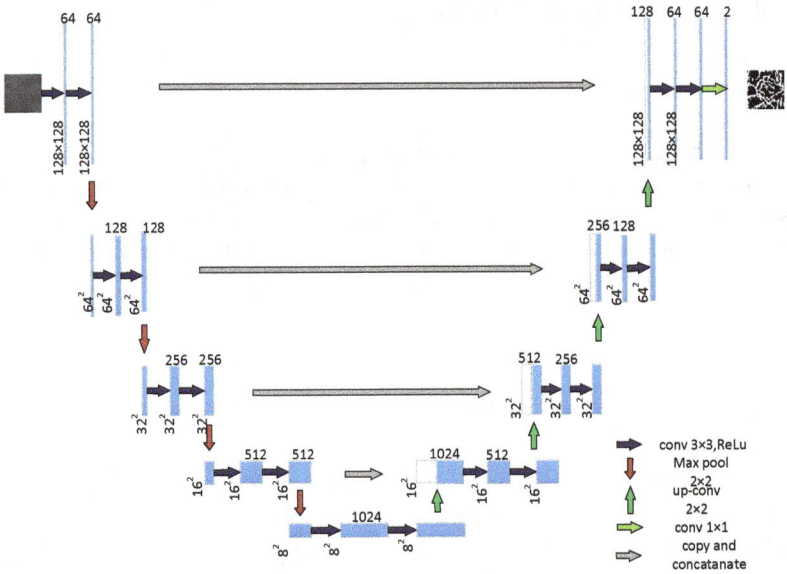

Fig. 2. The architecture of U-Net.

there are two 3×3 convolution layers, followed by a Leaky ReLU activation function $f(x) = max\ (x, leak \times x)$ and a 2×2 max pooling layer with stride 2. Here, the first derivative of Leaky ReLU activation function is 1 for $x > 0$ and $leak$ for $x \leq 0$. So, the leaky rectifier allows for a small, non-zero gradient when the saturated units and not active. In other words, the Leaky ReLU network still learns slowly when training traditional ReLU networks with constant 0 gradients. The number of convolution kernels in the next layer is twice times than the number of convolution kernels in the previous layer, with the number of convolution kernels in the fifth layer being 1024. In the expansive path, each layer has two 2×2 up-convolution layers, followed by two 3×3 convolution layers and a Leaky ReLU activation function.

Discriminator Architecture. The discriminator network has a typical CNN architecture that takes the input image of size 128 × 128 and outputs one decision: is this a real pair (ground truth) or is it a fake pair (output of generator)? In this network, there are four convolutional layers with a kernel size of 3×3 and one fully connection layer. Strided convolutions are applied to each convolution layer to reduce spatial dimensionality instead of using pooling layers. Batch-normalization is applied to each layer of the network, except for the input and output layers. The Leaky ReLU activation function is applied to all layers except the output layer which uses the Sigmoid function for the likelihood probability score of the image.

Objective Function. Let the generator G be a mapping from a hand image x to a vein image y. Then, taking a pair of (x, y) as input, the discriminator D makes a binary decision $\{0,1\}$, where 0 or 1 represent that y is produced by the generator or is the ground truth, respectively. Adversarial networks are trained by optimizing the following loss function of a two-player minimax game.

$$L_{adv} = \mathbb{E}_{x,y \sim p_{data}(x,y)}[logD(x,y)] + \mathbb{E}_{x \sim p_{data}(x)}[\log (1 - D(x, G(x)))] \quad (2)$$

where $\mathbb{E}_{x,y \sim p_{data}(x,y)}$ is the expectation over the pairs (x, y) sampled from the joint data distribution of real pairs $p_{data}(x, y)$ and $\mathbb{E}_{x \sim p_{data}(x)}$ is the expectation over the x sampled from the real vein network distribution $p_{data(x)}$.

Although optimizing the above loss function induces G to generate visually sharp results, recent work in [28] and [29] has shown that considering some global loss such as L_1 provides more consistent results for image synthesis. Inspired by works [31] and [32], as the binary cross entropy loss [34] has shown good performance for image segmentation, we employ it to penalize the distance between the ground truth and the output of generator and formulate accordingly the following objective function.

$$L_{seg} = L_{adv}(G, D) + \lambda L_{binary} \quad (3)$$

where $L_{binary} = \mathbb{E}_{x,y \sim p_{data}(x,y)} - y.\log(G(x)) - (1-y)[\log(1-G(x)]$ and λ balances the contribution of the two losses. To train the model, we use the Adam optimizer with a fixed learning rate of $2e^{-4}$ and the exponential decay rate for the 1st moment estimates of 0.5 and set the trade-off coefficient in Eq. (3) to 500 ($\lambda = 500$).

Vein segmentation. For GAN training, the input is two pairs of images, i.e. a synthetic pair and the associated real pair, as shown in Fig. 1. The generator aims then at producing the vein pattern to fool the discriminator while the discriminator tries to distinguish the real pair from fake pair (output of generator). After training, taking an image either from a training or a test image as input, the generator outputs a vein image map providing the probability of each pixel to belong to a vein pattern. This map allows us to perform segmentation, i.e. to obtain the vein patterns by a threshold of 0.5.

4   Experiment and Results

Our methods are implemented in Matlab and conducted on a high performance computer with 8 Core E3- 1270v3 3.5 GHz processor, 16GB of RAM, and a NVIDIA Quadro GTX1080ti graphics card. To test the performance of our approach, we have carried out several experiments on a public hand-vein database. We have employed three state of the art techniques, namely Repeated line tracking [15], Gabor filters [7], and Hessian phase [9], to segment the hand-vein network from each input image. Using Eq. 1, the resulting segmented image is used to label the vein pixels in the input image. In this way, a dataset is constructed to train the proposed model. For testing, our model takes a hand image as input and outputs an image map of the same size that is then subject to binarization using a threshold of 0.5. We match the resulting binary vein image for verification. In our tests, we compare our approach with the Maximum principle curvature [20] and CNN [25] techniques, which also show good performance for vein verification. The performances are shown in the following experiments.

4.1   CASIA Dataset

The CASIA Multi-Spectral Palm-print Image Dataset includes 7200 palm images, acquired from 100 different people. The images, collected by a self-designed multiple spectral imaging device, are taken from six different wavelength bands, which are 460 nm, 630 nm, 700 nm, 850 nm, 940 nm and white light respectively, and have a resolution of 8 bit. All images are captured in two separate sessions with a time interval of more than one month. In each session, the left and right hands of each subject provide 3 image samples respectively. We only employ the images acquired under the 850 nm wavelength. In total, there are 1200 images (100 subjects × 2 hands × 2 sessions × 3 images). In our experiments, a pre-processing method [26] is employed to extract the region of interest (ROI) image and to resize the resulting image to 128×128.

4.2   PolyU Dataset

The PolyU Palm Image Dataset consists of 3000 palm images that are collected from 250 subjects in two separate sessions with a time interval of 9 days. All images are captured under four illuminations, namely blue, green, red and near-infrared (NIR). In each session, the left and right hands of each subject provide 12 image samples respectively, under each illumination. 6000 NIR images (12 NIR images × 2 hands × 250 subjects) are employed to test our approach as we perform vein verification. We employ a pre-processing method [26] to segment the region of interest (ROI) image and normalize it to the spatial resolution of 128×128.

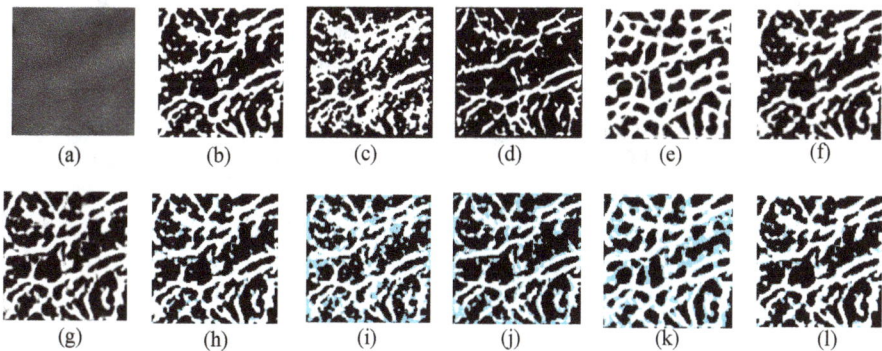

Fig. 3. Segmented results. (a) Original hand-vein image; (b) vein patterns extracted from (a) using Hessian phase; (c) vein patterns extracted from (a) using Repeated line tracking; (d) vein patterns extracted from (a) using Maximum principle curvature; (e) vein patterns extracted from (a) using Gabor filter; (f) vein patterns extracted from (a) using CNN; (g) vein patterns extracted from (a) using the proposed approach; (h) difference (colored region) between (b) and (g); (i) difference (colored region) between (c) and (g); (j) difference (color region) between (d) and (g); (k) difference (color region) between (e) and (g); and (l) difference (color region) between (f) and (g).

4.3   Verification Results

The CASIA Dataset (A) consists of 200 hands. In our experiments, we split it into three datasets: 70 hands with 420 (70 hands × 6 samples) samples for the training dataset (A1), 30 hands with 180 (30 hands × 6 samples) samples for the validation dataset (A2), and 100 hands with 600 (100 hands × 6 samples) samples for the test dataset (A3). On the other side, dataset B consists of 500 palms associated with 6000 images. Similarly, there are 2400 (200 hands ×12 images) images for the training set (B1), 600 (50 hands ×12 images) images for the validation set (B2), and 3000 (250 hands ×12 images) images for the testing set (B3). The training dataset is used to train our approach while the validation dataset is employed for hyper-parameter selection. After training, our model can be used to segment the vein network. For comparison, we also apply existing some approaches to extract the vein patterns. The segmentation results of the compared approaches are shown in Fig. 3 (a)-Fig. 3 (g). To facilitate the comparison, Fig. 3 (h)-Fig. 3 (l) show the difference between the vein patterns extracted by the proposed approach and those extracted by the other approaches. From Fig. 3 (a)-Fig. 3 (g), we observe that the Repeated line tracking-based and the Gabor filter-based segmentation approaches are prone to over-segmentation (generate additional non-vein patterns). By contrast, the Maximum principle curvature segmentation technique fails to extract some actual vein patterns. Hessian phase and CNN show relatively better performance, but these techniques generate noise and broken vein patterns. Compared to existing approaches, the proposed approach is robust to noise and can extract connected and a smooth vein network, as shown in Fig. 3 (g). Also, as we can see from Fig. 3 (h)-Fig. 3 (l), Hessian phase and CNN achieve similar segmentation results to our approach while there is large difference between the vein patterns extracted by our approach and those extracted by the remaining approaches, i.e. Maximum principle curvature, Repeated line tracking and Gabor filter.

Table 1. EER of various approaches on CASIA database

Methods	CASIA data-base	PolyU data-base	Computation time (s)
Repeated line tracking [15]	4.00	2.96	8.35
Maximum principle curvature [20]	2.33	3.87	1.43
Gabor filters [7]	1.00	0.93	2.96
Hessian phase [9]	1.33	0.78	1.25
CNN[25]	0.74	0.06	5.58
The proposed approach	0.33	0.03	0.03

From the quantitative viewpoint, our experiments aim at assessing the verification performance of our approach w.r.t existing approaches. For the CASIA dataset, the test set A3 consists of 100 hands with 600 (100 hands × 6 samples) samples, captured in two sessions. We select the 3 hand-vein images acquired during the first session as training data while the corresponding 3 images acquired during the second session are employed as testing data. Thus, 300 (100 × 3) genuine scores are produced by matching images from the same hand. The impostor matching score computation is time costly as there are 178200 (6 × 6 × 100 × 99 / 2) matching groups. To reduce time cost, all hands are randomly split into 10 groups and then the impostor matching scores are computed for each group. For example, we divide the 100 hands from the test dataset into 10 groups where each group includes 60 (6 × 10) images from 10 hands. For each group, matching the *i*-th sample at different sessions from different hands (*i* = 1, 2, 3, 4, 5, 6) produces 270 (10 × 9×3) impostor matching scores. We obtain, in this way, 2, 700 (270 × 10 groups) matching scores for the 10 groups of the test dataset. For the PolyU dataset, there are 3000 (250 hands ×12 images) images. Similarly, matching the vein patterns from the same hand produces 16500 (250 × 12 × 11 / 2) genuine scores. Likewise, 36000 (10 × 12 × 25 × 24 / 2) impostor matching scores are generated by splitting 250 hands into 10 groups. The False Rejection Rate (FRR) is computed according to the genuine scores while the False Acceptance Rate is computed according to the impostor scores. The Equal Error Rate (EER) is the error rate when FAR is equal to FRR. The experimental results obtained with the various approaches mentioned previously on both datasets are listed in Table 1 and the corresponding receiver operating characteristics (ROC) curves (the FAR against the FAR) are depicted in Fig. 4.

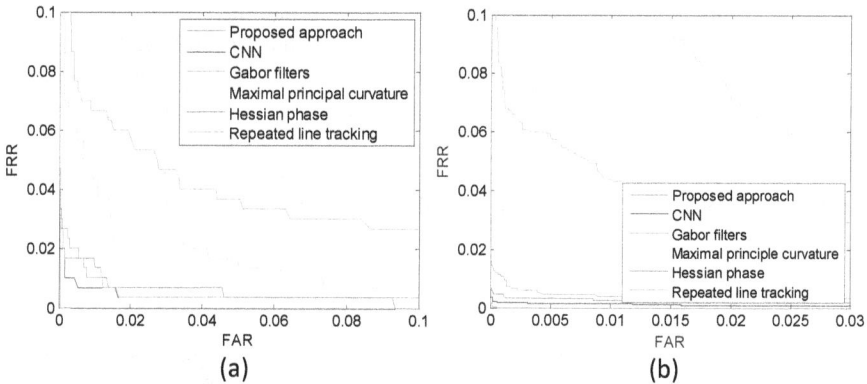

Fig. 4. Receiver operating characteristics on (a) CASIA database and (b) PolyU database.

The experimental results (Table 1 and Fig. 4) show that the proposed approach outperforms the existing approaches in terms of reducing the verification error and the average computation cost. The Repeated line tracking, Gabor filters and Hessian phase methods achieve 4.00% EER, 1.00% EER and 1.33 EER, respectively, on the CASIA dataset. Our GAN-based model brings a dramatic improvement by reducing the verification error to 0.33%. For the PolyU dataset, a similar trend is observed: the proposed approach achieves a much lower EER than the remaining approaches. Overall, the deep learning based approaches (i.e. CNN and our GAN-based approach) achieve lower EER w.r.t the handcrafted features engineering based segmentation approaches. This can be explained by the fact that the latter extract explicitly some image processing-based features (low level features) that might discard relevant information for vein pattern classification. By contrast, the deep learning based methods automatically learn high-level features that are directly related to vein patterns, through the loss function and back-propagation. We also observe that the proposed model achieves a significantly lower EER than CNN. Such a good performance may be attributed to the fact that the CNN takes a patch as input. For small patches, more detailed vein patterns are extracted but including more noise. This noise can generate mismatch errors, which reduces verification accuracy. On the contrary, if patches are too large, the CNN will take into account more global information, which results in missing detailed vein textures and inaccurate localization. This is because the CNN classify only the central pixel of the patch. Unlike CNN, the proposed approach takes the entire image as input instead of patch, which enables it to consider the localization and global context at the same time, thereby achieving higher verification accuracy. Unlike the CNN, our GAN-based approach classifies all the image pixels in one pass, and, thus, does not suffer from the issues mentioned above related to the small or large size of the patches.

As shown in Table 1, we also compare the proposed GAN model with existing approaches in terms of inference time. It can be observed that the repeated line tracking method has the highest time cost for single image segmentation, followed by the CNN model. Compared to the existing methods, including CNN, the proposed approach achieve by far the lowest time cost (i.e. 0.03s) to segment a hand-vein image. This may be explained by the fact that the proposed approach takes the entire image as input instead of patches. It can predict, therefore, the vein patterns by classifying all the image pixels in one forward pass, which substantially reduces the computation time.

5   Conclusions

In this chapter, we have presented a new GAN model to extract palm veins. First, the vein pixels are labeled by the combination of several handcraft-based segmentation methods. Second, unlike traditional mathematical models with an *a priori* assumption on the data distribution, our GAN neural network allows us, through its loss function, to model the underlying distribution of plausible images directly from the training data without any distribution assumption. Accordingly, we have built a GAN based vein segmentation model which consists of generator and discriminator for this purpose. The Generator is a U-Net that transforms vein images into vein patterns. The discriminator is a convolutional neural network with four convolutional layers and a fully connected layer. The proposed model is trained adversarially and the U-Net is employed to predict the probability of the palm vein pixels. To perform matching, the output of the U-Net is subject to binarization by thresholding. Finally, we match the resulting binary images for verification. Our experimental results on two large public datasets demonstrate the ability of the proposed approach to improve performance dramatically, by significantly reducing the verification error and computation time w.r.t the state of the art.

References

1. A. Jain, L. Hong, and R. Bolle: On-line fingerprint verification, IEEE Transactions on Pattern Analysis and Machine Intelligence, 19(4), 302–314 (1997).
2. J. Daugman: How iris recognition works. IEEE Transactions on Circuits and Systems for Video Technology 14(1), 21–30 (2004).
3. Han J, Bhanu B: Individual recognition using gait energy image. IEEE transactions on pattern analysis and machine intelligence 28(2), 316–322 (2005).
4. M. A. Turk and A. P. Pentland: Face recognition using eigenfaces. CVPR, 586–591, 1991.
5. A. Kumar and D. Zhang: Personal recognition using hand shape and texture. IEEE Transactions on Image Processing 15(8), 2454 (2006).
6. Bromley J, Guyon I, LeCun Y, et al: Signature verification using a siamese time delay neural network. Advances in neural information processing systems, pp: 737-744 (1994).
7. A. Kumar and Y. Zhou, "Human identification using finger images," IEEE Transactions on Image Processing, vol. 21, no. 4, pp. 2228–2244, 2012.
8. Y. Zhou and A. Kumar, "Human identification using palm-vein images," IEEE Transactions on Information Forensics and Security, vol. 6, no. 4, pp. 1259–1274, 2011.
9. A. Kumar and K. V. Prathyusha, "Personal authentication using hand vein triangulation and knuckle shape," IEEE Transactions on Image Processing, vol. 18, no. 9, pp. 2127–2136, 2009.
10. V. Ruizalbacete, P. Tomegonzalez, F. Alonsofernandez, J. Galbally, J. Fierrez, and J. Ortegagarcia, "Direct attacks using fake images in iris verification," in Biometrics and Identity Management, First European Workshop, BIOID 2008, Roskilde, Denmark, May 7–9, 2008. Revised Selected Papers, 2008, pp. 181–190.
11. Menotti D, Chiachia G, Pinto A, et al. Deep representations for iris, face, and fingerprint spoofing detection[J]. IEEE Transactions on Information Forensics and Security, 2015, 10(4): 864–879.
12. T. Tanaka and N. Kubo, "Biometric authentication by hand vein patterns," in Sice 2004 Conference, 2004, pp. 249–253 vol. 1.
13. A. K. Jain, A. Ross, and S. Prabhakar, "An introduction to biometric recognition," IEEE Transactions on Circuits Systems for Video Technology, vol. 14, no. 1, pp. 4–20, 2004.
14. M. Shahin, A. Badawi, and M. Kamel, "Biometric authentication using fast correlation of near infrared hand vein patterns," International Journal of Biomedical Sciences, vol. 2, no. 1-4, pp. 141–148, 2007.
15. N. Miura, A. Nagasaka, and T. Miyatake, "Feature extraction of finger-vein patterns based on repeated line tracking and its application to personal identification," Machine Vision and Applications, vol. 15, no. 4, pp. 194–203, 2004.
16. H. Qin, L. Qin, and C. Yu, "Region growth–based feature extraction method for finger-vein recognition," Optical Engineering, vol. 50, no. 5, pp. 057 208–057 208, 2011.
17. H. Qin, X. He, X. Yao, and H. Li, "Finger-vein verification based on the curvature in radon space," Expert Systems with Applications, vol. 82, pp.151–161, 2017.
18. L. Yang, G. Yang, Y. Yin, and X. Xi, "Finger vein recognition with anatomy structure analysis," IEEE Transactions on Circuits and Systems for Video Technology, 2017.
19. H. Qin, L. Qin, L. Xue, X. He, C. Yu, and X. Liang, "Finger-vein verification based on multi-features fusion," Sensors, vol. 13, no. 11, pp. 15 048–15 067, 2013.

20. Miura N, Nagasaka A, Miyatake T. Extraction of finger-vein patterns using maximum curvature points in image profiles[J]. IEICE TRANSACTIONS on Information and Systems, 2007, 90(8): 1185–1194.

21. C.-B. Yu, H.-F. Qin, Y.-Z. Cui, and X.-Q. Hu, "Finger-vein image recognition combining modified hausdorff distance with minutiae feature matching," Interdisciplinary Sciences: Computational Life Sciences, vol. 1, no. 4, pp. 280–289, 2009.

22. S. Chaudhuri, S. Chatterjee, N. Katz, M. Nelson, and M. Goldbaum, "Detection of blood vessels in retinal images using two-dimensional matched filters," IEEE Transactions on Medical Imaging, vol. 8, no. 3, pp. 263–269, 1989.

23. B. Huang, Y. Dai, R. Li, D. Tang, and W. Li, "Finger-vein authentication based on wide line detector and pattern normalization," ICPR, pp. 1269–1272, 2010.

24. Z. Zhang, S. Ma, and X. Han, "Multiscale feature extraction of finger-vein patterns based on curvelets and local interconnection structure neural network," ICPR, vol. 4, pp. 145–148, 2006.

25. Qin H, El-Yacoubi M A. Deep representation-based feature extraction and recovering for finger-vein verification J. IEEE Transactions on Information Forensics and Security, 12(8): 1816–1829 (2017).

26. Qin H, El Yacoubi M A, Lin J, et al. An Iterative Deep Neural Network for Hand-Vein Verification. IEEE Access, 2019, 7: 34823–34837.

27. Yang W, Hui C, Chen Z, et al. FV-GAN: Finger Vein Representation Using Generative Adversarial Networks[J]. IEEE Transactions on Information Forensics and Security, 2019, 14(9): 2512–2524.

28. Huafeng Qin, Mounîm A. El-Yacoubi. Finger-vein quality assessment by representation learning from binary image. Neural Information Processing. Springer International Publishing, 2015, 10.1007/978-3-319-26532-2_46.

29. Qin H., El-Yacoubi M. A., Finger-Vein Quality Assessment Based on Deep Features From Grayscale and Binary Images. International Journal of Pattern Recognition and Artificial Intelligence, 33(11), 2019.

30. Qin H., El-Yacoubi M. A., Deep Representation for Finger-Vein Image-Quality Assessment[J]. Circuits and Systems for Video Technology, IEEE Transactions on, 2018, 28(8):1677–1693.

31. Zhang, Y., Yang, L., Chen, J., Fredericksen, M., Hughes, D.P., Chen, D.Z., 2017c. Deep adversarial networks for biomedical image segmentation utilizing unannotated images, in: International Conference on Medical Image Computing and Computer-Assisted Intervention, Springer. pp. 408–416.

32. Dou, Q., Ouyang, C., Chen, C., Chen, H., Heng, P.A., 2018. Unsupervised cross-modality domain adaptation of convnets for biomedical image segmentations with adversarial loss. arXiv preprint arXiv:1804.10916.

33. I. Goodfellow, J. Pouget-Abadie, M. Mirza, B. Xu, D. Warde-Farley, S. Ozair, A. Courville, and Y. Bengio, "Generative adversarial nets," in Advances in neural information processing systems, 2014, pp. 2672–2680.

34. Ronneberger O, Fischer P, Brox T. U-net: Convolutional networks for biomedical image segmentation[C]//International Conference on Medical image computing and computer-assisted intervention. Springer, Cham, 2015: 234–241.

Chapter 4

Diabetic Retinopathy Analysis Based on Retinal Fundus Photographs via Deep Learning

Yufang Tang[1,2,*], Yan Xu[1,2]

[1]Shandong Normal University, Jinan, Shandong 250014, China
[2]Concordia University, Montreal, Quebec H3G 1M8, Canada
*tangyufang@outlook.com

Abstract: As deep learning methods are being used increasingly in the field of retinal image analysis, the challenges of insufficient and imbalanced dataset begin to affect the performance of the retinal image analysis models. In this chapter, we introduce a training method which uses data augmentation to overcome the insufficiency and imbalance of retinal image dataset, particularly the retinal image for the classification of diabetic retinopathy (DR). By using the data augmentation method, we are able to obtain a balanced augmented retinal image dataset from the original imbalanced retinal image dataset to expand the size of the original retinal dataset about four times, leading to better classification accuracy.

Keywords: Diabetic retinopathy analysis, Retinal fundus photographs, Deep learning, Feature map, Data augmentation

1 Introduction

1.1 *Harm of Diabetic Retinopathy*

DR is the most common cause of preventable blindness, mainly affecting working age population in the world [1], it's caused by damage to the blood vessels of the light-sensitive tissue at the back of the eye (retina). It is a complication of both types of diabetes mellitus, which affects the light perception part of the retina.

People might not have symptoms in the early stages of DR. As time goes on, DR symptoms may appear as shown in Figure 1[1], and diabetic retinopathy usually affects both eyes.

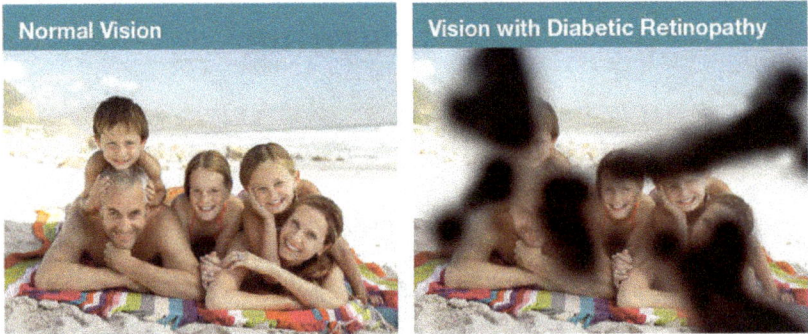

Fig. 1: Symptoms of diabetic retinopathy

DR has became an urgent issue to the global health system as the number of people with DR and visual-threatening diabetic retinopathy (VTDR) are both growing [2]. Thus, it is an important task to develop an intelligent system for retinal disease detection.

1.2 *Methods for Diagnosing Diabetic Retinopathy*

Regular eye exams, good control of your blood sugar and blood pressure, and early intervention for vision problems can help prevent severe vision loss.

To diagnose DR, the physicians need to analyze the fundus images of the patients. The fundus photographs are visual records which demonstrate the current ophthalmoscopic appearance of patients' retina as well as the characteristics of diabetic retinopathy [3]. Thus, it is significant for the physicians to analyse the fundus images when diagnosing DR.

Nowadays, there are two major methods to diagnose DR. The first one is to manually inspect the morphological changes in microaneurysms, exu-

---

[1] https://www.natural-health-news.com/diabetic-retinopathy-causes-symptoms-diagnosis-and-treatment/

dates, blood vessels, hemorrhages, and macula in the retinal image, which is time-consuming and tedious [4].

The elimination of avoidable blindness is critically dependent on the pool of well-qualified ophthalmologists and supporting eye care infrastructures [5]. Recent research has given better understanding of requirement in clinical eye care practice to identify better and cheaper ways of identification, management, diagnosis and treatment of retinal disease.

The other method is to employ computer-aided diagnosis (CAD), powered by deep learning, and combined with human intervention, which has better performance and efficiency in detecting DR.

Computer-aided disease diagnosis in retinal image analysis could ease mass screening of population with diabetes mellitus and help clinicians in utilizing their time more efficiently.

The diagnosis of retinal disease based on retinal image analysis is very important to people's health, especially in preserving people's vision, since DR is one of the most common cause for blindness [6]. This makes it perfect for development and evaluation of image analysis algorithms for early DR detection.

The retinal disease diagnosis powered by deep learning has great potential in applications including DR detection, since the deep network is able to exploit both simple and complex compositional features of retinal disease images [7]. Moreover, the deep network does not rely on hard-coded knowledge, but its self-learning ability to recognize raw data [7], using deep learning network can save a tremendous amount of human resources and achieve higher diagnosis accuracy.

Although the deep learning techniques for retinal images have surpassed most of the state-of-the-art-methods [7], there are still challenges remaining. As we know that training a good model may require a massive and balanced dataset of labeled retinal images [7], which is difficult to acquire. Thus, it becomes necessary to develop a method to generate more retinal images by using data augmentation methods, so that we can obtain more and balanced dataset used for training.

2 Acknowledgement of Diabetic Retinopathy

2.1 *Pathogenesis of Diabetic Retinopathy*

Over time, too much sugar in the blood can lead to the blockage of the tiny blood vessels that nourish the retina, cutting off its blood supply. As a result, the eye attempts to grow new blood vessels. But those new vessels do not develop properly and can leak easily, as shown in Figure 2[2].

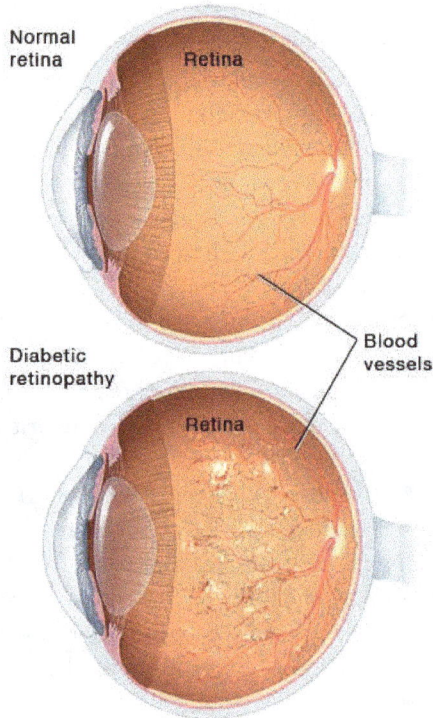

Fig. 2: Normal retina vs. diabetic retinopathy

There are two types of diabetic retinopathy, One is early diabetic retinopathy, called nonproliferative diabetic retinopathy (NPDR) in the most common form, and new blood vessels are not growing (proliferating).

Diabetic retinopathy can progress to more severe type, known as proliferative diabetic retinopathy (PDR) in Figure 3. In this type, damaged blood vessels close off, causing the growth of new, abnormal blood vessels in the retina, and can leak into the clear, jelly-like substance that fills the center of the eyes (vitreous).

---

[2]https://www.mayoclinic.org/diseases-conditions/diabetic-retinopathy/symptoms-causes/syc-20371611

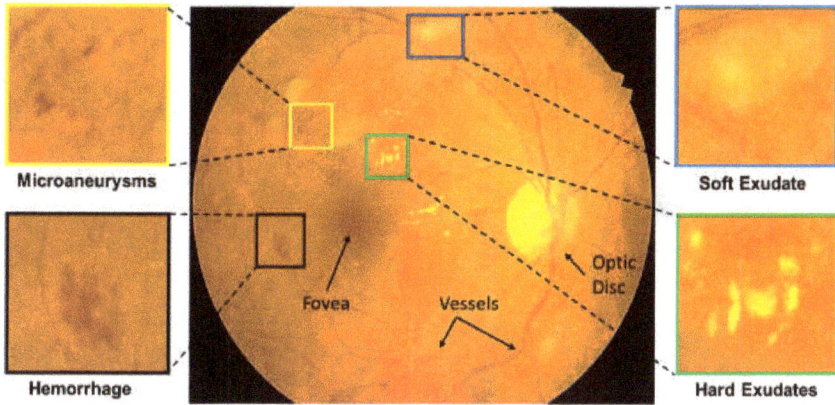

Fig. 3: Four several retinal lesions

2.2 *Retinal Images*

The retina can be photographed directly by a fundus camera, which is a specialized microscope with a camera attached [3]. A retinal image contains a large amount of information, including characteristics of DR, such as macular edema and microaneurysms [3].

The basic structure of a normal human retina is shown in Figure 4[3]. The fovea is a dark round shape on the left side of the image, the macula is a round margin surrounding the fovea, and the optic disc is a bright round shape covered under the blood vessels.

The thick blue and red vessels in the center respectively refer to the retinal vein and artery, and the thin blue and red vessels that spread across the entire retina are respectively the retinal venules and arterioles.

2.3 *Retinal Diseases*

There are several diseases that manifest as retinal impairments, including DR, age related macular degeneration (AMD), cardiovascular diseases, glaucoma, and so on [7]. We mainly focus on improving the deep learning methods for DR diagnosis. DR is a disease serving as one of the prevalent consequences of diabetes mellitus (DM), caused by fluctuating level of glycemia.

When DR is severe, it will become microaneurysms, neovascularization, hemorrhages, cotton wool spots, and exudates in the retinal region. In

---

[3]https://www.visioncareofmaine.com/retina-bangor/

HUMAN EYE ANATOMY
THE RETINA

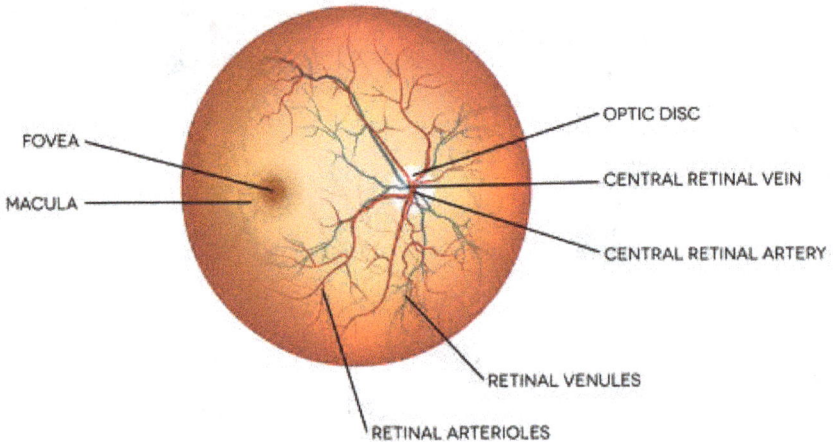

Fig. 4: The structure and components of a normal human retina

Figure 3, we can see four zoomed lesions.

The first one, the sac-like structure [7], enclosed by the yellow box is called microaneurysms, which refers to unusual dilation of retinal capillaries caused by lack of oxygen supply [7]. The emanation of fluffy white pitches [7] enclosed by blue box is called soft exudates or cotton wool spots. The lesion enclosed by the black box is called hemorrhages, which refers to the burst of vessels [7] caused by the blockade in arterioles. The part enclosed by the green box is called hard exudates, which refers to a lesion caused by leakage of fat, protein and water from the walls of retinal vessels [7].

3   Diabetic Retinopathy Analysis Based on Deep Learning

3.1   *Deep Learning*

Deep learning is a subfield of machine learning, and they both employ neural network. Deep learning allows computational models that are composed of multiple processing layers to learn representations of data with multiple levels of abstraction.

The role of data analysis in health informatics has grown rapidly in the last decade. Data in a deep learning network is transmitted from one neuron to another one, and this kind of non-deep machine learning requires manual data representation in the form of the features [7]. This means if the maiden machine learning is used, then we need to "teach" the machine about the features of all possible lesions for DR detection.

A deep learning machine, with its ability to conduct hierarchical feature extraction [7], has self-learning ability and does not require handcrafted features to be fed.

3.2 *Our Methods for Fundus Disease Diagnosis*

We use four models: Model 1[4], Model 2[5], Model 3[6] and Model 4[7] with different complexities to verify our hypothesis about the impact of balancing the dataset and expanding the size of the dataset.

For example, the structure of Model 1, which has been described as a mimic of the VGG16 neural network, is shown in Figure 5[8].

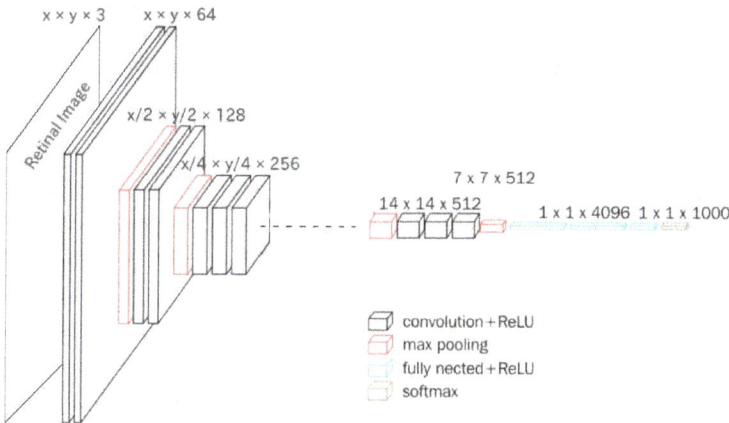

Fig. 5: The architecture of VGG16

---

[4]https://www.pyimagesearch.com/2018/05/07/multi-label-classification-with-keras/
[5]https://machinelearningmastery.com/how-to-develop-a-convolutional-neural-network-to-classify-photos-of-dogs-and-cats/
[6]https://towardsdatascience.com/image-classification-python-keras-tutorial-kaggle-challenge-45a6332a58b8
[7]https://blog.keras.io/building-powerful-image-classification-models-using-very-little-data.html
[8]https://neurohive.io/en/popular-networks/vgg16/

VGG16 is a convolutional neural network (CNN) model proposed in paper [8]. The model achieves 92.7% top-5 test accuracy in ImageNet. It was one of the famous models submitted to ILSVRC-2014. It makes improvement over AlexNet by replacing large kernel-sized filters (11 and 5 in the first and second convolutional layers, respectively) with multiple 3x3 kernel-sized filters one after another.

The Model 1 contains five main structures: the first one is an input module, the second and third are convolutional modules, the fourth is a flattening module, and the last is a softmax output module.

The first module contains a Con2D layer, a "relu" activation layer, a batch normalization layer, a maxpooling layer and a dropout layer. The second and third modules both have two groups of Con2D, "relu" activation, batch normalization, then a maxpooling and a dropout. The fourth module has a flattening layer, a dense layer, a "relu" activation layer, a batch normalization layer and a dropout layer. The last module has a dense layer and a "softmax" activation layer. For the other three models, they both have a simpler structure than Model 1, and use simple layers, including Con2D, "relu" activation, maxpooling and dropout layers.

3.3   *Balance of Training Dataset*

Deep learning is supervised learning, so the training dataset is very important. Generally, it is common knowledge that too little training data results in poor approximation. An over-constrained model will underfit the small training dataset, whereas an under-constrained model, in turn, will likely overfit the training data, both resulting in poor performance. Too little test data will result in an optimistic and high variance estimation of model performance.[9] Research has indicated that, the increase of data imbalance will have a negative effect on the performance of classification models [9].

In the experiment mentioned above, the data with different class prevalence indexes were fed into the same group of neural network models, and the result shows that as the class prevalence index gets closer to 0.5, which represents a balanced class prevalence, the performance of the model increases, and eventually leads to the conclusion about the impact of imbalance dataset on the performance of neural network models.

Moreover, research has also shown that increasing the size of dataset can lead to better accuracy and improved generalization capability [10]. In

---

[9]https://machinelearningmastery.com/impact-of-dataset-size-on-deep-learning-model-skill-and-performance-estimates/

their another experiment, three datasets with the size of 400, 800, 1200 were fed into the same training model with the same data partition ratio. The results show that the error of the models to process unknown data decreases as the size of the dataset increases, which lead to the conclusion of the impact of dataset size mentioned above. Therefore, by analysing those two experiments conducted, we can learn that it is very beneficial to create a balanced dataset with a considerable size to achieve satisfying performance.

3.4   *Data Augmentation*

Deep CNN has performed remarkably well on many computer vision tasks. However, these networks are heavily reliant on big data. Unfortunately, many application domains do not have access to big data, such as medical image analysis. Data augmentation is a data-space solution to the problem of limited data. [11]

Data augmentation encompasses a suite of techniques that can enhance the size and quality of training dataset so that better deep learning models can be built using the new dataset. The image augmentation algorithms include geometric transformation, color space augmentation, kernel filters, mixing image, random erasing, feature space augmentation, adversarial training, generative adversarial networks (GAN), neural style transfer, and meta-learning. [11]

4   Experiments

4.1   *Dataset*

The diabetic retinopathy detection dataset used by this experiment come from Kaggle[10], which is one of the largest data science communities in the world. It has five zip packages for training, five zip packages for testing, and a csv file for storing the ground truth of all ten packages, and we used the five training packages in our experiment.

Before training, we reorganize the file structure of the Kaggle dataset, by checking the csv file, we find out each image's label, write it on its filename, and divide all training and testing images into two separate folders.

The Kaggle dataset was classified into five severity degrees of diabetic retinopathy. Degree 0 corresponds to no DR, degree 1 corresponds to mild,

---

[10]https://www.kaggle.com/c/diabetic-retinopathy-detection/data

degree 2 corresponds to moderate, degree 3 corresponds to severe, degree 4 corresponds to proliferative DR. Due the objective conditions, such as the actual unbalance distribution of the number of patients in each degree, the Kaggle dataset is also very unbalanced.

We merge the images from the five training packages, obtain 13423 images for degree 0, 1312 images for degree 1, 2755 images for degree 2, 449 images for degree 3 and 384 images for degree 4, and we use this image dataset to train our group A (including Model 1, Model 2, Model 3 and Model 4).

4.2   *Task*

The task is to grade fundus images according to the severity level of diabetic retinopathy and diabetic macular edema.

There may be a presence of venous beading, retinal neovascularization which can be utilized to classify DR retinopathy into one of the two phases known as non-proliferative diabetic retinopathy (NPDR) and proliferative diabetic retinopathy (PDR) as shown in Figure 6 (a) and (b).

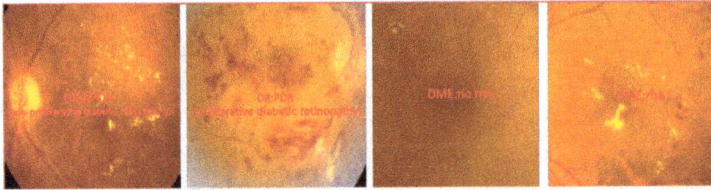

(a)

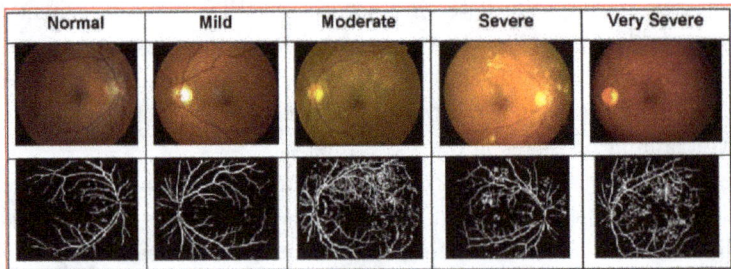

(b)

Fig. 6: (a) NPDR and (b) PDR

The determination of DR and DME severity are based on criteria given in Table 4.1. It is essential to decide the need for treatment and follow-up recommendations.

Table 4.1: Diabetic retinopathy severity scale

Level	Disease Severity	Findings
Grade-0	No apparent retinopathy	No visible sign of abnormalities
Grade-1	Mild-NPDR	Presence of Microaneurysms only
Grade-2	Moderate-NPDR	More than just microaneurysms but less than severe NPDR
Grade-3	Severe-NPDR	Any of the following: (1) >20 intraretinal hemorrhages, (2) Venous beading, (3) intraretinal microvascular abnormalities (4) No signs of PDR
Grade-4	PDR	Either or both of the following: (1) Neovascularization, (2) Vitreous/pre-retinal hemorrhage

4.3 Data Augmentation

In an effort to expand and balance our dataset, In this chapter, we employ several augmentation methods to produce augmented images for training. These augmentation methods include rotation, width and height shifting, rescaling, zooming, horizontally and vertically flipping, brightness and filling. We compare with two pairs of original and augmented images, from Figure 7 (a) to (d).

After the augmentation process, we get 13423 original images for degree 0, 13092 augmented images for degree 1, 13244 augmented images for degree 2, 13235 augmented images for degree 3, 13216 augmented images for degree 4, as shown in Table 4.2, and this image dataset is used to train group B (including Model 1, Model 2, Model 3 and Model 4). Note that the degree 0 images are original, since there are already sufficient images to make the entire dataset generally balanced.

4.4 Accuracy and Loss

From the four pairs of Figure 8, (a) and (b), (c) and (d), (e) and (f), (g) and (h), show the four models' performance when training with the original dataset and the augmented dataset.

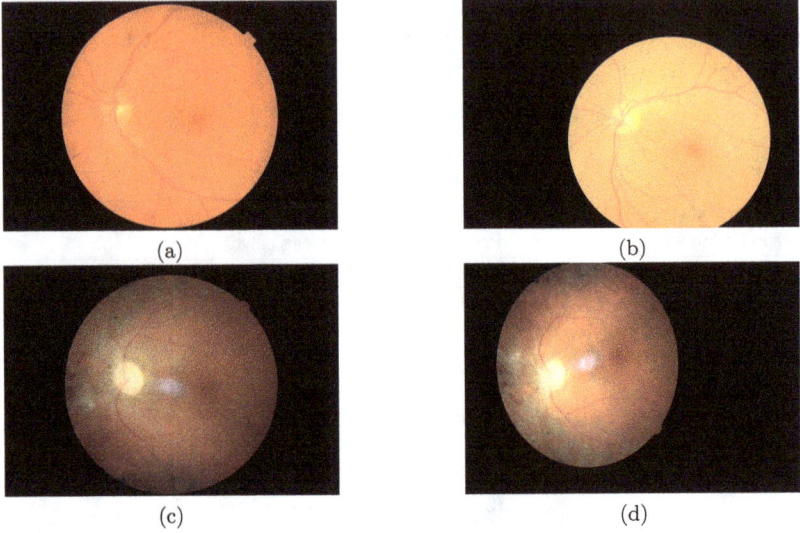

(a)

(b)

(c)

(d)

Fig. 7: Two pairs of original and augmented retinal images. (a) Original retinal image (degree 0). (b) Augmented retinal image (degree 0). (c) Original retinal image (degree 4). (d) Augmented retinal image (degree 4)

Table 4.2: The number of original and augmented retinal images with different degrees

Degree	Number of original retinal images	Number of augmented retinal images
0	13423	13423
1	1312	13092
2	2755	13244
3	449	13235
4	384	13216

It can be seen that, when considering the scaling of y-axis, Model 2, Model 3 and Model 4 displayed improvement in accuracy when training with the augmented dataset.

The accuracies of Model 2, Model 3 and Model 4 are normally below 0.5 when training with the original dataset, while the accuracy of the same model was above 0.5 when training with the augmented dataset.

For Model 1, training with the augmented data does not have lower accuracy than that with the original data. Considering all the above results, we can conclude that the data augmentation can bring a certain degree of

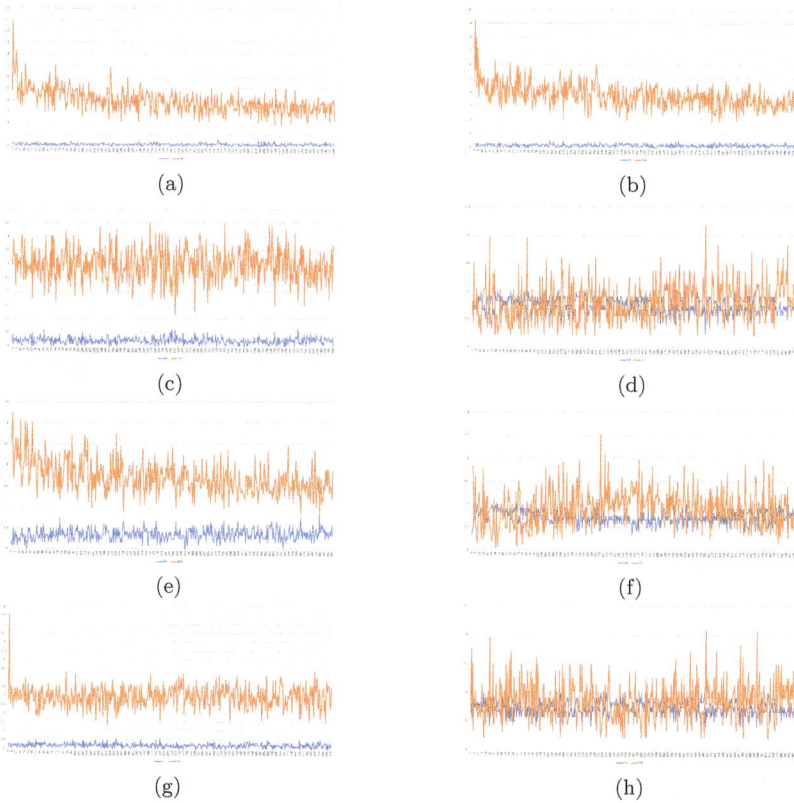

Fig. 8: Two pairs of original and augmented retinal images. (a) Model 1 trained using original data. (b) Model 1 trained using augmented data. (c) Model 2 trained using original data. (d) Model 2 trained using augmented data. (e) Model 3 trained using original data. (f) Model 3 trained using augmented data. (g) Model 4 trained using original data. (h) Model 4 trained using augmented data.

improvement on the accuracy of retinal disease diagnosis models.

On the other hand, we manually calculate the accuracy of our model. 505 images from the validation set, which have not been seen during training, are used as the prediction samples.

In Figure 9, we can observe that the Group B using the augmented dataset displays improved prediction accuracy. The accuracies are calculated as the total number of predictions that match the ground truth divided by the total number of images tested. We can see from Figure 9 that all the prediction accuracies of the models trained with the augmented dataset are better than the models trained with the original dataset.

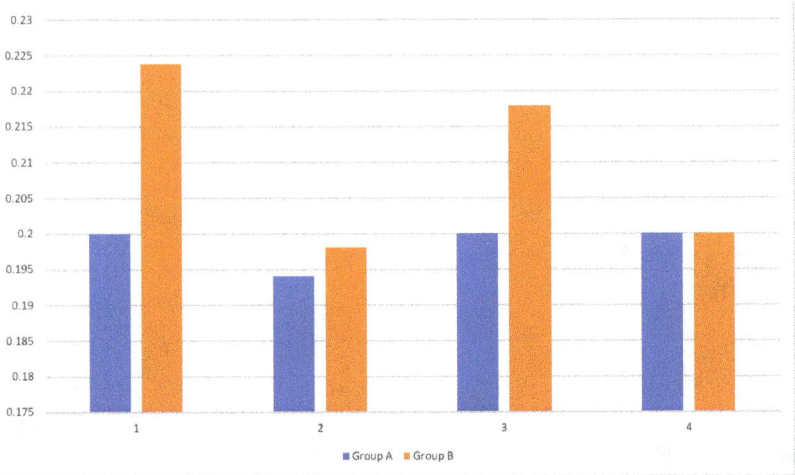

Fig. 9: Prediction results: Group A uses the original dataset, Group B uses the augmented dataset.

4.5 *Analysis*

We observe that the reason why data augmentation can improve the accuracy of the retinal disease model. That is because by using data augmentation, more training examples can be provided, which overcomes the issue of unbalanced dataset.

As mentioned above, the original dataset is extremely unbalanced. By using augmentation method, we can create an expanded and balanced retinal image dataset.

The augmentation method greatly increased the number of images for label 1, label 2, label 3 and label 4, as well as the overall size of the retinal image dataset.

5 Conclusion

Diabetic retinopathy (DR) is the most common cause of preventable blindness, mainly affecting working age population in the world, early intervention for vision problems can help prevent severe vision loss. In this chapter, we propose a method to improve the performance of grading DR based on deep learning.

The challenges of insufficient and imbalanced dataset affect the performance of deep learning, when we use deep learning to diagnose DR based on retinal images.

We use data augmentation to compensate the insufficiency and imbalance of retinal image dataset, obtain an balanced augmented retinal image dataset from a imbalanced original retinal image dataset, and get better classification accuracy.

From the experiment and analysis above, we can conclude that data augmentation can improve the accuracy of retinal disease diagnosis by deep learning.

References

[1] N. Cho, J. Shaw, S. Karuranga, Y. Huang, J. da Rocha Fernandes, A. Ohlrogge and B. Malanda, Idf diabetes atlas: Global estimates of diabetes prevalence for 2017 and projections for 2045, *Diabetes research and clinical practice* **138**, pp. 271–281 (2018).

[2] Y. Zheng, M. He and N. Congdon, The worldwide epidemic of diabetic retinopathy, *Indian journal of ophthalmology* **60**, 5, p. 428 (2012).

[3] J. Shankle, Ophthalmic photography: retinal photography, angiography, and electronic imaging, *Survey of Ophthalmology* **49**, 2, p. 264 (2004).

[4] J. Amin, M. Sharif and M. Yasmin, A review on recent developments for detection of diabetic retinopathy, *Scientifica* **2016**, pp. 1–20 (2016).

[5] D. Worsley and D. Simmons, Diabetic retinopathy and public health, *Automated Image Detection of Retinal Pathology* **2010**, pp. 27–66 (2010).

[6] Y. Zhang and M. An, An active learning classifier for further reducing diabetic retinopathy screening system cost, *Computational and mathematical methods in medicine* **2016**, pp. 1–10 (2016).

[7] M. Badar, M. Haris and A. Fatima, Application of deep learning for retinal image analysis: A review, *Computer Science Review* **35**, pp. 1–18 (2020).

[8] K. Simonyan and A. Zisserman, Very deep convolutional networks for large-scale image recognition, *arXiv preprint arXiv:1409.1556* (2014).

[9] M. A. Mazurowski, P. A. Habas, J. M. Zurada, J. Y. Lo, J. A. Baker and G. D. Tourassi, Training neural network classifiers for medical decision making: The effects of imbalanced datasets on classification performance, *Neural networks* **21**, 2-3, pp. 427–436 (2008).

[10] A. Ajiboye, R. Abdullah-Arshah and Q. Hongwu, Evaluating the effect of dataset size on predictive model using supervised learning technique, *International Journal of Software Engineering & Computer Sciences (IJSECS)* **1**, pp. 75–84 (2015).

[11] C. Shorten and T. M. Khoshgoftaar, A survey on image data augmentation for deep learning, *Journal of Big Data* **6**, 1, p. 60 (2019).

Chapter 5

Pattern Recognition in Handwriting

Sheila R. Lowe, MS, CFDE

Sheila Lowe & Associates, Write Choice Ink
sheila@sheilalowe.com

Abstract: This article discusses handwriting as a means of personality assessment. Through a discussion of the various internal and external factors, we see how the elements of spatial arrangement, writing form, and writing movement develop handwriting patterns. These are shown to combine in ways that demonstrate a wide range of personality traits in handwriting. The question is addressed whether cursive writing is still important in a digital age.

Keywords: Handwriting, Cursive, Graphology, Personality, Behavior

Introduction

For most people, handwriting is viewed as a series of lines of ink lying static on the page. With proper eye-training, it becomes a dynamic trail whose patterns reveal important information about the writer. This chapter discusses how the writing trail is developed, as well as the influences that shape it.

Note: The masculine gender is used for ease, but all genders are to be considered included in 'he.'

1   Factors that Affect Handwriting

Before we can begin our discussion of the writing trail and the patterns it produces, we must first consider the several and varied factors, both external and internal, which affect the production and resulting appearance of handwriting. There are cases where, without explanation of these factors, even the trained observer is apt to misinterpret what is seen on the page and thus, skew the results of an examination [1, p. 175].

As an example, Fig. 1 is the handwriting of a sixty-year-old man who worked at a Porsche mechanic. The writing patterns are quite interesting, but note that the words themselves are largely illegible. Since the purpose of handwriting is communication, it should be legible. Then how do we interpret such writing patterns?

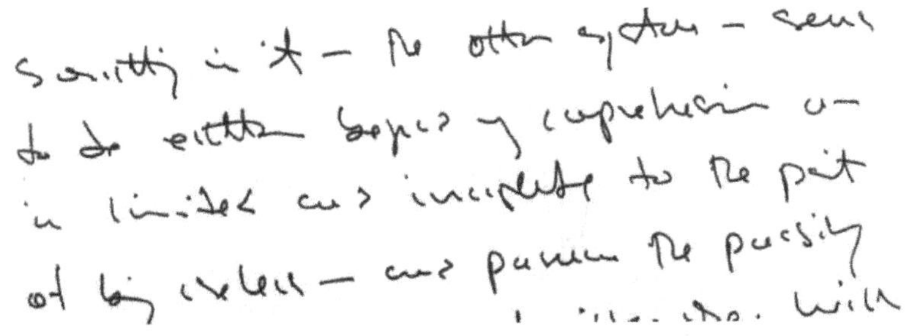

Fig. 1. Illegible Handwriting of an Autistic Man

In order to properly assess this writer, it would be of great importance for the observer to know that he is autistic and his autism affects his ability to communicate, both verbally and in handwriting (in typed writing he is clear and expressive). Without that information, the one making the assessment would very likely draw some incorrect conclusions.

1.1 External Factors

External factors include the most basic elements involved in handwriting production. These are the materials the writer chooses to use—the type of paper, the writing instrument, writing surface, writing position. Even the general writing environment can affect the way one writes.

Slick paper creates a different look than porous paper, as does the type of pen the writer selects. A fountain pen produces a different output than a ballpoint. Width of nib or penpoint and ink choice add more layers of complication. The writer accustomed to using an ultrafine point may be uncomfortable and even cranky when forced to use a much point, and vice-versa. As a result, the handwriting will have a different appearance than normal.

Even the decision of whether to choose pencil or pen will have an effect on how the writing looks. When using a pencil, a newly-sharpened point will produce a different result than a stubby point, which creates a blurry look.

Writing position, too, makes a difference. Holding a pad on one's knees while writing in bed, for example, or against someone's back while standing, rather than writing while seated at a desk makes a difference.

Environmental influences should also be considered. If you were to attempt to write while sitting in a helicopter, you could expect the movement of the vehicle to influence your arm and hand, and produce shaky results. One observing such a writing without knowing the circumstances under which it was done might lead him to believe the writer was suffering from some type of illness.

Guided hand

In cases of serious illness an individual might be requested to sign a document but be too weak to do so or otherwise be unable to hold the pen by himself. Under those circumstances, a second person may be asked to hold the ill person's hand and guide the signature. This is another instance where it would be important for one assessing the handwriting to know that the second person's own writing patterns and habits will influence the signature and alter the pictorial quality—the writing pattern.

1.2 Internal Factors

Internal factors affecting handwriting include the writer's age, general state of physical, mental, and emotional health, and medications or other chemical substances, as well as personality traits that develop over time [2].

Handwriting *may* change over time with advancing age, but this is not always the case and much depends on the writer's state of health. The handwriting of an eighty-year-old who suffers from Parkinson's disease will be affected by micrographia (reduced size) and tremor (shakiness in the writing trail), and affected further by medication. The handwriting of a healthy, active eighty-year-old is far less likely to show the effects of aging.

Chemical substances

Certain medications may affect handwriting patterns, some in positive ways, others negatively. Therefore, if a handwriting sample has some strange quality, it would be important to ask whether the person was taking some prescribed medication. The effects of marijuana, cocaine, heroin and other 'recreational rugs' are likewise seen in handwriting.

Alcohol exerts a strong influence on handwriting. A long-ago House of Seagram advertisement demonstrated this very effectively. The subject was asked to write the phrase *'I can drive when I drink'* when sober and again after each of a series of seven drinks. The handwriting started to deteriorate fairly rapidly.

Personality

Personality is considered an internal factor as it relates to handwriting and plays a major role in the evolution of handwriting patterns. Although personality continues to develop over time and can change with major life experiences, the basic character structure is generally set fairly early in childhood. Psychologists suggest that the maturation process continues through to the ages of 20 to 40.

More on personality traits later. First, we need to explore the genesis and development of handwriting skills in schoolchildren.

2   What Happened to Handwriting Training?

The requirement to teach cursive (joined-up) handwriting was dropped from the Common Core Curriculum in the United States, so that since around 2009, handwriting started to be referred to as a 'lost art.' Since then, a body of research in the field of education and neurology has demonstrated that the dearth of handwriting training is detrimental to brain development in young children.

In 2018, the American Handwriting Analysis Foundation published a white paper detailing more than eighty recent peer-reviewed studies that point to the efficacy of handwriting training. The white paper, titled *The Truth About Handwriting: Why it matters in a digital age,* is available in seven languages as a free <u>download</u>. Since the paper's publication, research in the area of handwriting and neurology has accumulated at a such a rapid pace that an update is already being compiled.

Since the removal of the requirement to teach handwriting, many states have come to recognize the problems attendant with leaving handwriting training out and have returned it to their curriculum.

A recent informal survey of the departments of education for all fifty states revealed that by spring of 2020, twenty-five required cursive training in public schools. An additional six states had legislation pending. Seven states left it to the individual school districts to decide.

Despite the many benefits stated in the numerous peer-reviewed studies published in reputable journals, ten states (AK, CO, HI, IN, MO, NB, ND, NH, VT, WA) still have no requirement to teach their students handwriting.

3   Printing and Writing Patterns

To develop a signature—a symbol of who you are, and an important personal projection of self to the world—one must be able to write in cursive, or joined up letters. A Harvard Law Review article [3] studied the signatures of CEOs and found that those with "larger signatures, an indicator of narcissism, perform worse…" than CEOs with small signatures. This is just one facet of one's signature and personality.

Thus, one of the many unfortunate side-effects of not providing children with this important skill is that they are unable to sign their name. A recent Stanford University study lamented that one problem in young people being unable to use cursive writing arises when they need to 'sign' their ballot when voting and how it "doesn't make matching signatures easier."

These students may learn to print-write, and as with cursive, there are many individualized forms of printed writing. Like any other graphic expression, printing can be analyzed for personality traits and character, but the writing patterns are generally less fluid and fluent. Printing literally breaks the bridges from one letter to the next and is symbolic of breaking contact between people.

How many people do you see walking around (or at the dinner table), nose in their phone or tablet, sending text messages rather than engaging in interpersonal face-to-face contact? The cultural impact of breaking bonds between people is reflected in the broken patterns in handwriting.

The good news is that, as mentioned above, while handwriting training virtually disappeared in the twenty-first century, it is now coming back.

With the effects of the global pandemic of Covid-19 affecting schooling, exactly what will be in the curriculum remains to be seen. What we do know is, as more states see the importance of providing children with this vital lifelong skill, the continuing need for training in handwriting pattern recognition and interpretation will continue to grow.

Parents of children who do not attend a school that teaches handwriting but want them to learn can avail themselves of one of the good instruction systems such as Cursive Logic to train them at home, or find one of the many 'cursive coaches' to work with them.

Most children love to write in cursive. They say it makes them feel like grownups and helps them write faster. They like the way it looks. Added to that, research indicates that brain benefits include better spelling, greater ability to retain more important information than what is absorbed by merely keyboarding, and many others.

3.1 Methods of Handwriting Instruction

In U.S. schools that require handwriting training, instruction most often begins in the third grade. The student is first taught to write using a specific copybook writing method.

For most U.S. students who learned to write before the late 1970's their schools followed the Palmer or Zaner-Bloser methods, though some Southern states used others. In 1978, the D'Nealian copybook was introduced, which simplified cursive writing somewhat over the older, more flowery methods.

In the twenty-first century several new methods have been introduced, including the New American Cursive and Cursive Logic copybooks, which are highly recommended by professionals in the handwriting field.

Regardless of the copybook taught, the general writing techniques are the same. The student is first taught to form the individual letters of the alphabet. Once he is comfortable writing the abc's, he can be said to be writing in *letter impulse* which requires *finger movement*. In other words, the fingers are the focus and the hand makes small movements.

The next step is to learn to string the individual letters together into whole words. This is *word impulse*, which uses more of the hand and requires greater *wrist movement.*

Finally, after a great deal of practice, when the student has learned to form words into whole sentences, he is said to be writing in *sentence impulse* and is using *arm movement,* which allows freer range of motion. Depending on personality type, the child will unconsciously choose the type of movement that fits best with his temperament. A shy, retiring child will be more likely to stay with the finger movement, while an exuberant, outgoing child will use the larger arm movement.

When the student has practiced enough that he can write without having to stop and think about how to form each individual stroke, each letter and each word, he has reached a level of *graphic maturity*. In other words, he knows how to write.

3.2 Individualization

Once the student is familiar with the learned copybook style and has attained graphic maturity, he begins to individualize his handwriting. Once again, personality plays a role in the individualization process.

A student who prefers to stick closely to what he knows and is good at following direction rather than striking out on his own may continue to write in the style of the learned copybook model throughout his life. As adults, such people often are drawn to a career in fields such as administration, teaching, or other areas where they are expected to toe the mark and follow the rules. Thus, very little, if any individualization is displayed in their handwriting. It continues to follow the copybook model.

Those who do individualize their handwriting show independence and a need to be their own person. The individualization process is accomplished by one of two general processes: simplification or elaboration, which we will discuss next.

Note: the following information applies to adult handwriting as well as children's.

Simplification

Simplification refers to the elimination of handwriting features that are unnecessary to maintain legibility. We might view this as akin to removing the training wheels on a bicycle once a child has learned to ride.

Those extra features, which includes upper and lower loops and beginning and ending strokes, were needed as a crutch to help in the learning process. With the development of graphic maturity, these are no longer needed and can safely be eliminated.

The writer who chooses to simplify his handwriting demonstrates personality traits that are unlike one who adheres to copybook, or one who elaborates his writing.

Beginning with simplification, there is a range of how much simplification the writer uses. Depending on other factors that are beyond the scope of this discussion, moderately simplified writing suggests self-confidence and quick thinking.

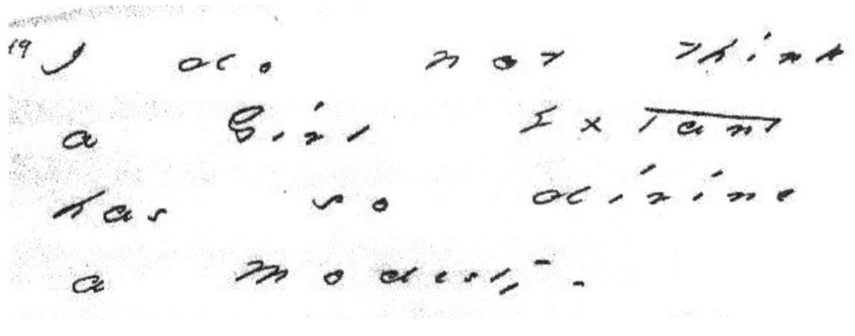

Fig. 2. Poet Emily Dickinson

It should be noted that *legibility* is a key to whether the patterns created by a simplified writing are interpreted in a positive or negative manner. Writing that is simplified to an extreme and interferes with legibility may signify a vision problem, or in some cases, emotional or mental illness.

Legibility can be determined by taking a single word out of context to see if it can be read. The highly simplified handwriting in Fig. 2 is that of American poet Emily Dickinson.

Elaboration

Elaboration, which is more common in the handwriting of teens and adults than young children, is found in additions to handwriting that make it showy and possibly even ostentatious. Elaborate is seen in the addition of unnecessary strokes, extra-large or overly embellished capital letters, extremely tall or long and wide loops. In fact, any extremes that draw the eye fall under this category.

While self-confidence is generally a factor in one who simplifies, chances are that the writer who greatly elaborates his handwriting is not receiving positive feedback and modeling at home and feels prompted to find some means to say "Hey! Look at me! I exist!" Without intervention, the behavior can become problematic. Depending on the type and degree

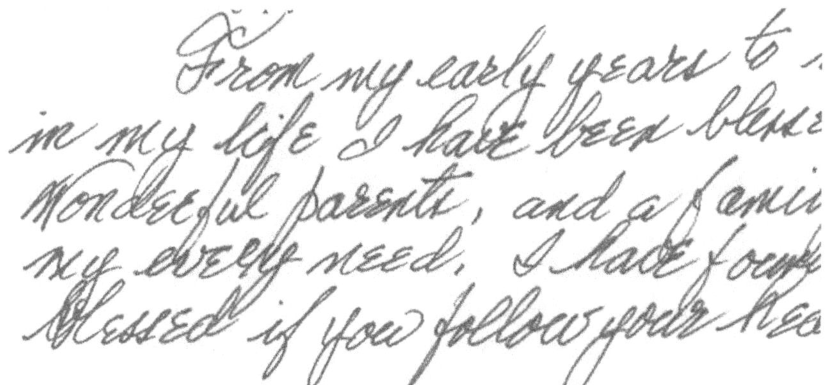

Fig. 3. Elaborate writing

of elaboration, the writer who chooses to individualize in this way tends to draw attention to himself in other ways, too.

A moderate degree of elaboration suggests the ability to step into the spotlight and feel comfortable being seen. *Overly* elaborated writing may have an abundance of curlicues, bloated loops, or extreme size, and often indicates pathology that requires counseling. Fig. 3 is an example of elaborate handwriting in an adult.

4   Master Patterns

Having reached graphic maturity, the writer creates his own *master pattern* for each letter. That is, any letter of the alphabet might be written in more than one form. If the letter 'a,' for example, appears at the beginning of a word, it may be formed differently than when it appears in the middle, or at the end of a word. Those various styles of a single letter comprise the writer's master pattern for that letter. The same is true for each letter of the alphabet.

Some features of handwriting, notably size and slant, are affected by mood, and change on a temporary basis. More permanent changes occur to some degree over time, which reflect the writer's response to major life experiences. For the most part, though, a writer's master patterns will remain stable. And that is why, when an expert is provided with enough exemplars to demonstrate an individual's writing habits, with an understanding of pattern recognition, handwriting can be authenticated.

5   Dimensions in Handwriting

Some readers may be surprised to learn that handwriting exists in multiple dimensions. In English, the pen progresses across the page from left to right. In Hebrew or Arabic, it moves from right to left. Some forms of Asian writing can be written from right to left in vertical lines. Yet, no matter which direction the writing flows, the number of dimensions—an integral part of the pattern of writing movement—are consistent.

The first dimension is the movement of the pen from one side of the page to the other. The second is the up and down movement which forms the upper and lower letters as the writing advances.

Pen pressure, which moves into and out of the paper, creates depth and is a third dimension. Movement *above* the paper from the time the pen is lifted from the page and follows an invisible trail in the air until it is set down once again, is known as an 'airstroke,' and creates the fourth dimension.

6   Unified Patterns

A unified pattern that cannot be properly identified by simply describing the sum of its parts is known as the 'gestalt.' Applying this concept to handwriting, a mere description of the size, the slant, the pressure, and other discrete parts of a specific handwriting sample, is not very useful, even if personality traits are assigned to each of these features.

On the other hand, observing handwriting patterns in terms of the arrangement, form, and movement as a whole provides a complete picture that allows the sum to be greater than its parts.

6.1 Figure-Ground

Part of gestalt theory is the perceptual concept of figure-ground. In the context of handwriting, the *ground* is the paper and the *figure* is the written trace upon the ground.

It can be said that a 'good gestalt' is when the figure-ground relationship is in balance. In a page of handwriting this means nothing needs to be altered to make it look better, the overall arrangement is balanced.

What does *balance* in handwriting mean? It may be easier to describe *imbalance* first.

Unbalanced handwriting may, at one extreme, appear to have too much ink (figure) covering the writing material (ground). In other words, the page is covered with the writing trace and not enough white space is allowed to show through.

At the other end of the spectrum is too much white space (ground), with small amounts of writing (figure) floating between large islands of paper. The writing looks lost on the page.

As in all parts of handwriting, the personality of the writer will determine how the figure/ground proportions are laid out and how they should be interpreted.

7   Space–Form-Movement

Handwriting patterns can be described in terms of three basic pictures, all of which are interconnected to create the gestalt: spatial arrangement, writing form, and writing movement.

It is the interplay of these three pictures that creates the writing patterns and allows for personality analysis as well as handwriting authentication. (Our focus for this article is on the personality aspects.)

The various configurations of space, form, and movement are seen along a continuum of strong, weak, or disturbed, which together demonstrate the writer's personality traits and behaviors.

It is impossible to correctly interpret the patterns in each of these pictures without keeping in mind the whole—the gestalt. Why? Because according to Kurt Koffka, one of the founders of gestalt theory, the whole is 'other than the sum of its parts' [4].

Always keeping in mind that no piece or part of handwriting exists in a vacuum, we explore these basic pictures in the next section.

7.1 Space

Space is comprised of the layout or overall arrangement of the writing on the page and is the most unconscious element of handwriting.

Space provides a setting for the ink and represents the writer's environment. Place a penpoint on a blank sheet of paper and all you have produced is a dot. But once the hand begins to move, forms are created in the writing trail and a pattern soon begins to emerge.

The number of times the writing trail stops and restarts, the sequence and intersection of lines, the heights to which the trail rises, its horizontal expansion, its depth (how deeply the pen presses into the paper), the proportions, the degree of connectivity and types of ligatures that bind the letters to each other, along with hundreds of other variables, work together to create patterns that represent the individual whose hand moves the pen.

Research has shown that the countless types of handwriting patterns demonstrate aspects of personality and behavior [5].

7.2 Form

Form is the most conscious aspect of handwriting. It is the *style* that the writer chooses to represent himself and causes the writing to look the way it does. Writing form is equivalent to the font choice one makes in a writing a computer document.

Form is created by the writing movement and affected by it. Both form and movement impact the space.

We have already discussed some of the most basic writing forms or styles: copybook, simplified, elaborate. To these we can add the numerous types of printed writing, and 'printscript,' which combines printing and cursive in varying degrees.

7.3 Movement

Movement makes the greatest impact on the entire picture. It's what draws the eye and lets the viewer know instinctively whether the writing has a good gestalt (is balanced on the page). That first dot placed on the page is the departure point. Without movement, the dot remains static; there will be nothing more.

When the pen begins to move, things happen [6].

Movement carries the flow of ink from one point to the next. It reveals how the writing line progresses and regresses along its path. It is a point of departure that activates an instinctive pattern and takes the writer where he wishes to go, and that may be in any direction, whether it is appropriate or not.

Appropriate?

What might not be appropriate in handwriting? Think of capital letters, as an example. A student is taught to begin writing capitals at the baseline of writing. The stroke then rises before it returns to the baseline.

After the individualization process some writers will start or finish their capital letters far below the baseline where they do not belong. The 'inappropriate' movement, in this case is below the baseline. To those who understand how personality impacts handwriting, this is a revealing feature of the writing pattern.

The handwriting movement reveals information about the amount of energy a writer has, as well as how and where it is distributed. The handwriting movement of a dynamic, active person is quite different from one who is quiet and passive. Look at the following two extremes in Fig. 4 and Fig. 5. It is easy to see which is more active and which is more passive.

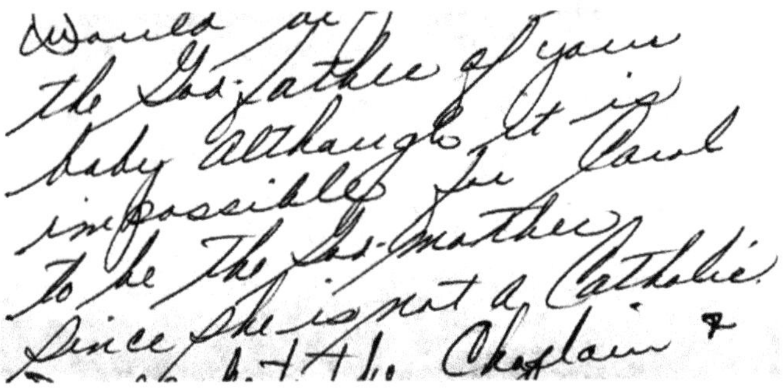

Fig. 4. Active Writing

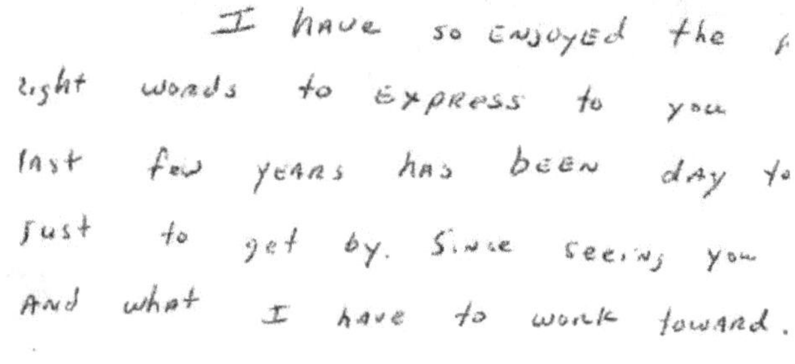

Fig. 5. Passive Writing

Form Consciousness/Persona

A *natural* pattern of movement has a balanced ebb and flow, but in the end, it must go forward. Psychiatrist Carl Gustav Jung wrote:

> "If one does not constantly walk forward, the past sucks one back. The past is like an enormous sucking wind that sucks one back all the time. If you don't go forward you regress. You have constantly to carry the torch of the new light forward so to speak, historically and also in your own life. As soon as you begin to look backward sadly, or even scornfully, it has you again. The past is a tremendous power." [7]

We look at what we have done in order to use our experiences to help us in the future. That is what writing movement does. The copybook cursive letter l, for example, starts at the baseline, the writing trail moves up, makes a turn to the left, returns to the baseline and ends to the right. It moves backwards then forward.

Natural movement in handwriting does not draw attention to itself with overblown loops or excessive size or other exaggerations and elaborations that bring the eye to a specific feature. It has a free-flowing, pleasing appearance that reflects an effortless fluency and a likewise natural and uncontrived personality with a strong (but not excessive) ego.

The handwriting in Fig. 6 demonstrates a reasonably good balance between contraction and release.

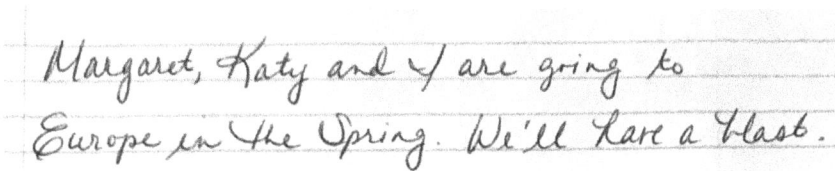

Fig. 6. Balanced contraction and release

7.4 Persona Writing Patterns

'Persona' is the mask or masks one shows in public situations, which may be somewhat different from the private self. We call handwriting that is 'form-conscious' and looks 'perfect,' rather than natural, persona writing.

The use of a persona writing signifies an attempt to project greater strength than the writer feels inside. He hides behind a public mask that is intended to conceal what he believes to be the lesser parts of himself and is intended to conceal the true, unacceptable self. As someone put it, "public smiles, private tears."

Virtually all of us use more than one persona. For example, you might display a different part of the self when giving a lecture, or shopping at the market, or reading a story to your children. That is different from someone who has developed such a strong social mask that it's as if the mask has overtaken the person. They behave quite differently in public than in a close relationship. Of course, this is reflected in their handwriting.

The following series of handwritings in Fig. 7 display various levels of persona writing patterns. Persona writing may be attractive or artistic, but the point is, the emphasis is always on the form more than the movement

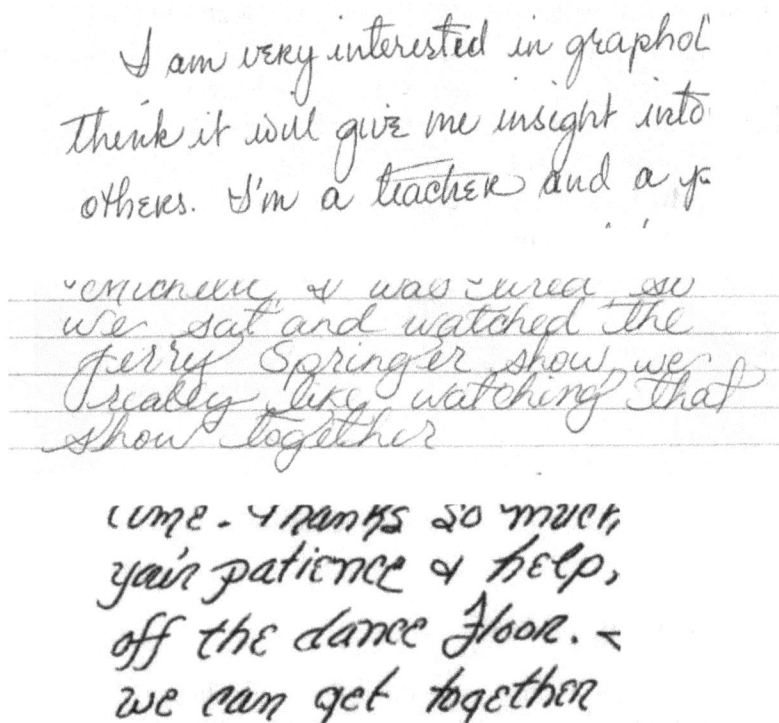

Fig. 7. Types of Persona Handwriting

and detracts from the natural appearance of the writing so that it looks more as if it was drawn [8].

8   The Contraction and Release Continuum

Grip on the writing instrument combined with way the writer uses the extensor and flexor muscles in the hand affects the degree of tension in the stroke and ultimately determines the rhythm of the writing pattern.

At one end of the contraction-release spectrum is the writer whose tight grip near the point of the pen barrel produces the tight, narrow, contracted handwriting of one who needs to always feel as in control of the environment as the pen.

At the other end of the spectrum is a loose grip with the pen held further back on the barrel. This produces more expanded writing with wider loops in one who is more impulsive and free-spirited.

Between the two extremes is a balance between contraction and release, with several gradations on each side.

9   A Brief History

In pre-WWII Europe, handwriting analysis, known as *graphology*, was well-researched and studied in major universities and was widely accepted as a personality assessment tool. When Hitler came to power, he outlawed it under the fortune telling laws and only allowed certain people he approved of to practice it.

For the next fifty years or so, handwriting analysis was studied quietly under wraps in most European countries except, perhaps, France, where it seemed to flourish for a time, and Israel, where it is apparently still popular. In recent years, the practice of graphology has come into favor in India, too.

In the United States, there have been mixed reactions since the war years when handwriting analysis was introduced by Louise Rice, an American newspaper woman. Without basis, or citing flawed studies, some psychologists have claimed it is an occult or pseudoscience. Other psychologists have found it an extremely useful tool in their practice.

It is unfortunate that there is no licensing of handwriting analysis, which rightly belongs to the behavioral sciences. The sad fact is, anyone, with no qualifications whatsoever, can call themselves a handwriting analyst and begin practice without regulation. Some practitioners have earned themselves a bad reputation, damaging their clients with bad information and unethical behavior, which has tainted the entire discipline.

Fortunately, ethical professionals have taken their studies seriously and have undergone rigorous training and testing. They attain certification through a reputable non-profit handwriting analysis organization, such as the <u>American Handwriting Analysis Foundation</u>. Their work in staffing, counseling, education, and other aspects of human interaction has kept handwriting analysis alive and well-respected through time.

10 In Summary

The patterns created by the flow of ink make handwriting like a fingerprint, as unique as the person who wrote it. And within those patterns, the variations of handwriting features and the way they interact number into the tens of thousands.

It is possible to make inferences about personality through graphic forms such as handwriting by studying the spatial arrangement, letter designs (form), and writing movement, all of which blend to demonstrate human behavior.

While handwriting professionals face the challenge presented by the Common Core Curriculum [9] having removed the requirement for public schools to teach handwriting, the states are beginning to recognize the damage that has been done to students.

Even though the rudimentary printed styles most children have been taught since the Common Core are not ideal for personal analysis, well-trained handwriting analysts have, through research, developed reliable methods to recognize personality traits that may be observed in those patterns. Just as with any other graphic movement, they can accurately describe the countless diverse and disparate personalities behind them.

References

[1] A. M. Headrick and R. A. Huber, Handwriting Identification: Facts & Fundamentals, Boca Raton, FL: CRC Press, 1999.

[2] A. Rafaeli and R. Klimoski, "Inferring personal qualities through handwriting analysis," *Journal of Occupational Psychology*, vol. 56, no. 3, pp. 191-202, 1983.

[3] N. Seybert, "Size Does Matter (in Signatures)," *Harvard Law Review,* 2013.

[4] B. Wong, "Gestalt principles (Part 1)," *Nature Methods*, vol. 7, no. 11, 2010.

[5] L. Fogarolo, "Graphic Pressure: How personality moves about the dual world," 2007. [Online]. Available: http://www.graphology.it/learning/graphology_course/graphic_pressure.

[6] S. Lowe, "Revealing Human Personality Through Handwriting," in *International Conference on Pattern Recognition and Artificial Intelligence*, Hong Kong, 2020.

[7] M.-L. v. Franz, The Cat: A Tale of Feminine Redemption (Studies in Jungian Psychology by Jungian Analysts #83), Inner City Books, 1999, p. 129.

[8] S. Lowe, "Persona Handwriting," (Zoom), 2017.

[9] "Common Core State Standards Initiative".

Chapter 6

A Deep Reinforcement Learning-based Study on Handwriting Difficulty Analysis

Chandranath Adak[1,2], Bidyut B. Chaudhuri[3,4], Michael Blumenstein[2]

[1]Centre for Data Science, JIS Institute of Advanced Studies & Research,
JIS University, India-700091
[2]Australian AI Institute, University of Technology Sydney, Australia-2007
[3]Techno India University, India-700091
[4]CVPR Unit, Indian Statistical Institute, India-700108
chandra@jisiasr.org

Abstract: A writing system contains several patterns/graphical symbols, which represent linguistic constructs. Such patterns/symbols are orthographically related to the phonology of a spoken language. A writing system may contain a few thousands of such symbols (e.g., about 10,000 frequently-used symbols for the Chinese writing system, about 1000 for Devanagari, etc.). Therefore, a section of people face difficulties in learning and writing all the character structures of a writing system. Besides, due to the dominance of electronic typing in this digital age, individuals often experience increasing difficulty in writing the characters by hand, even though they can read those. In this chapter, we analyze the handwriting of an individual and attempt to understand which character shapes are challenging for that person to write. We propose here a deep reinforcement learning-based model to predict the difficulty measure/score of a character written by an individual. For experimentation, we have chosen Bengali, a script of the Indian subcontinent, which contains about 1000 complex character shapes. We have obtained some promising results from our experiments.

Keywords: Character amnesia, Deep reinforcement Learning, Handwriting difficulty

1 Introduction

Handwriting is a cornerstone of human civilization, which preserves and passes the information on through the ages. Although handwriting is still valid in schools and some business work, the individual writing practice is declining in this digital era owing to the dominance of electro-mechanical typing/keyboard usage [1], born-digital documents [2], and paperless offices [3]. The handwriting of letters, manuscripts, diaries, applications, forms, cards, etc. is decreasing rapidly in today's cyber world. Due to such a decline in writing habits, difficulty in handwriting is observed by educational researchers and handwriting experts [4–6]. Sometimes, poor writing practice leads people to strain while writing a script by hand, although they can still read the material easily.

Handwriting difficulty arises even more for a writing system such as Chinese, Devanagari, Bengali, etc. [7], where thousands of various character shapes/symbols are present that take years to learn [8]. In such a script, often people struggle with writing characters due to *character amnesia* [4]. The *Romanization* of scripts is one of the plausible reasons behind it [9].

Difficulty in handwriting may also be perceived due to some illnesses such as Parkinson's, Alzheimer's disease, Dyslexia, Dysgraphia, Autism, Tourette syndrome, etc. [10–12].

In the educational and psychological research domains, studies on handwriting difficulty are frequently observed [5, 6, 13–15]. The reported past investigations are mostly based on case studies, and diagnoses are performed by educational practitioners and occupational therapists [5, 6, 13]. The practitioner analyzes multiple cognitive and linguistic factors involving writing difficulties [5]. Some researchers like [15] extracted dynamic features, e.g., peak count in a drawing velocity profile, mean drawing velocity, mean pen pressure, etc., from online hand-drawn data, and performed a basic quantitative kinematic analysis to detect children at risk of handwriting difficulties. In [6], a theory-based diagnosis for writing disabilities is reported, which was performed by some scientific practitioners. The writing difficulties in patients with Alzheimer's disease and mild cognitive impairment were examined in [12]. In this study, an analysis of variance (ANOVA) [12] was performed among various groups of writers.

The analysis of handwriting difficulty may be essential to investigate some educational, psychological, neurological, and cognitive problems of an individual [5, 6, 13]. After a proper handwriting difficulty analysis, *handwriting intervention* [14] can be performed by occupational therapists. Past investigations on handwriting difficulties mostly depend on case-based studies, diagnoses, and experiences of educational practitioners [5,13,14]. In this chapter, we attempt to impart the machine a human-like perception that understands handwriting difficulty, which may be an aid to occupational therapists for handwriting intervention. To the best of our knowledge, our research is the earliest attempt on handwriting difficulty analysis from the perspective of machine learning. Our preliminary work in this direction can be found in [16].

Some significant situations where our work is applicable are briefly mentioned below.

i) *Child's handwriting analysis*: During the development of handwriting competency, some writing difficulties can be observed among children. Such difficulties lead to diminished confidence and create distress among them [6]. Therefore, a system is essential to analyze the child's handwriting difficulty.

ii) *Adult's handwriting analysis*: In this digital world, due to declining writing habits, an adult may suffer from character amnesia and may exhibit consequent handwriting difficulties [4]. An automated system that can detect those characters, which are problematic to write by an individual, is beneficial. This system helps in self-assessment, so that people can practice more in writing problematic characters.

iii) *Healthcare*: Some diseases such as Parkinson's, Alzheimer's, Tourette's, Dyslexia, Dysgraphia, Autism, etc. may lead to handwriting difficulties [10–12]. Therefore, a system that inspects such difficulty may be useful in understanding the disease development and progression.

iv) *Aid to the occupational therapist*: Handwriting intervention is performed by the occupational therapist to tackle individual handwriting difficulty, and corresponding psychological and cognitive problems [5,6,13]. An automated handwriting difficulty analyzer may assist the practitioner in the intervention.

v) *Assistance in penmanship*: The handwriting difficulty analyzer may be employed as a self-evaluation tool by a person for learning/practicing penmanship and calligraphy [17].

In this chapter, we propose a deep reinforcement learning [18]-based model that generates handwriting difficulty scores with respect to the scribbled character images. Here, a reinforced agent focuses on various parts/fragments of the handwritten stroke, and predicts the difficulty measure using regression analysis [19]. The agent employs a CNN (Convolutional Neural Network) [20] for deep feature extraction from handwritten stroke, and uses an RNN (Recurrent Neural Network) [21] for memorizing the explored stroke fragments. A novel internal reward shaping scheme is also proposed for reinforcement learning. Our study focuses on handwriting difficulty understanding from an angle of machine learning, which is pioneering and opens up a new frontier of research. Moreover, from the perspective of the DAR (Document Analysis and Recognition) [22] community, handwriting difficulty analysis is a relatively new issue.

We work with offline Bengali handwriting in this chapter. Reasonably, the degree of people's handwriting literacy is worse for Bengali conjunct characters, which is more difficult to learn and remember owing to its orthographic complexity, compared to the basic characters [23]. More details on Bengali conjunct characters are presented in Section 2. The rest of the chapter is organized as follows. In Section 3, we formulate the problem being considered. The experimental data acquisition is described in Section 4. Subsequently, Section 5 discusses the proposed method, and Section 6 presents the experiments and results. Finally, Section 7 concludes the chapter with some ideas on the frontiers.

2 Briefings on Bengali Conjuncts

Bengali (endonym, *Bangla*) is an Indo-Aryan language [24], which is practiced by more than 265 million people, mostly hailed from south-east Asia. It is the national language of Bangladesh and one of the official languages of India. Apart from the Bengali language, the Bengali script is used to write some other languages, e.g., Assamese, Bishnupriya Manipuri, Hajong, Maithili, etc. Having more than 1000 years of a long root connected with Brahmi script [25], the modern handwritten/printed form of Bengali script came into existence in 19th century AD. Bengali script belongs to the Abugida class of scripts [23], where apart from individual vowel (e.g., আ) and consonant characters (e.g., ম), consonants modified by vowels (e.g., ম + আ = মা) are present.

Additionally, in Bengali script, up to four consonants can be conjoined to produce a consonant-conjunct which can further be merged with a vowel-modifier [23]. There are various forms of such conjunctions. In the *fused* form of conjunction, a stroke of a constituent consonant serves as a stroke of another (e.g., ণ + প = ঞ). In *approximate* form, two consonants are written very close by removing the usual character-gap (e.g., দ + গ = দগ).

The *compressed* form denotes where a consonant member of the conjunct character is squeezed (e.g., হ + হ = হ্ছ). In *abbreviated* form, a consonant member partially loses its individual shape (e.g., ঞ + জ = ঞ্জ). The *variant* form occurs when a consonant member changes its shape entirely (e.g., র + ক = র্ক). In the *altered* form of conjunction, all participating members fully lose their individual shapes and transform into a new shape (e.g., ঙ + গ = ঙ্গ). During the conjunction, the *sequence* of characters is a crucial factor; for example, a consonant র conjoined before and after another consonant ক, creates two separate conjuncts, র্ক (= র + ক) and ক্র (= ক + র), respectively. Moreover, some characters become structurally quite similar due to conjunction and may create *confusion*, e.g., ক্ষ (= ক + ষ) / হ্ম (= হ + ম), স্প (= স + প) / স্প (= স + প), এ / ঞ (= ঞ + র), ও / ত (= ত + ত) / ক্র (= ক + ত) / ৩ (= ণ + ড) / ৩ (= গ + ঐ), etc.

Contemporary Bengali script contains 50 basic character shapes, out of which 11 are vowels and the rest are consonants [23]. However, the conjunction of consonants creates about 300 new character structures. The consonants and conjunct-consonants also merge with vowel modifiers, which increases the size of the character set.

In recent times, due to some educational practitioners, while forming a conjunct, a constituent consonant is written vertically/horizontally touching to the other (e.g., ই, জ্ঞ). These are sometimes called *transparent* conjuncts against the classical *non-transparent* categories (e.g., ক্র, স্ম). The addition of such transparent conjuncts increases the Bengali character set size even larger. Besides, multiple typefaces (e.g., স্ম (= ক + স), ক্র (= ক + র) in *Nirmala UI* and স্ম (= ক + স), ক্র (= ক + র) in *Bangla Akademi* fonts), various cursive writing styles [26] (e.g., ঞ (= জ + জ), জ্ঞ (= জ + জ)), and writing idiosyncrasies [27] introduce many compound character-structures. Overall, about 1000 conjunct-character shapes are commonly found in the Bengali script.

3   Problem Formulation

Handwriting difficulty depends on human perception. In this work, our objective is to make the machine learn this perceptivity in a quantitative manner. Here, we define a handwriting difficulty measure, based on which it is possible to label the difficulty of a character written by an individual. To define this measure, we adopt the idea of subjective opinion score [28].

For ground-truthing, two different scores on a given handwritten character are considered in this work: *i)* expert opinion score and *ii)* self-assessment score. We now briefly discuss the details of both the scores.

i) Expert opinion score (A_μ): A set of handwriting experts first provide their opinion scores (A_i) on a given handwritten character within a continuous range of $[A_L, A_H]$; $A_L, A_H \in \mathbb{R}^+$. The aggregated expert opinion score is then computed in terms of the arithmetic mean (A_μ), which is defined as: $A_\mu = \sum\limits_{i=1}^{E} A_i$, where $E \geq 1$ is the number of experts who provided their subjective opinion scores on a handwritten character.

ii) Self-assessment score (A_s): Each writer provides a self-assessed difficulty score (A_s) to scribble a character within the same range of $[A_L, A_H]$.

For our task, we choose $A_L = 0$, $A_H = 10$, and $E \geq 30$.

Once we have both the scores from the experts and the writer, we compute an aggregated difficulty score on a given handwritten character. In this process, we first normalize both the scores as follows: $a_\mu = \frac{A_\mu - A_L}{A_H - A_L}$ and $a_s = \frac{A_s - A_L}{A_H - A_L}$. Here, $0 \leq a_\mu, a_s \leq 1$. We then take a weighted average to obtain the final difficulty score (a) corresponding to the given handwritten character, which is computed as $a = w_s.a_s + (1 - w_s).a_\mu$, where, w_s is a hyperparameter. Empirically, w_s is chosen as 0.15, which indicates that the lower weightage of the self-assessment score than the expert opinion score. It may be noted that $0 \leq a \leq 1$, where $a = 0$ and $a = 1$ refer to the lowest and highest difficulty scores, respectively. In other words, the difficulty score of a handwritten character closer to 1 implies the writer struggles highly to write that character.

We now have a set of character images and their actual difficulty scores, i.e., $(\mathcal{I}, a)$, with respect to the writers for the system-training.

Objective: The task is to predict the difficulty score $(\hat{a})$ of an unknown character image written by an individual.

We formulate this framework as a regression problem and propose a solution using deep reinforcement learning.

4　Experimental Dataset Generation

For our experimental analysis, we required a dataset containing handwritten samples and corresponding difficulty measure. We did not find any publicly available database containing such information. Therefore, we had to generate our own database.

For our study, we chose a total of 133 frequently used distinct Bengali conjunct characters. For the dataset generation, we gave the writers a form with the printed version of the conjuncts and boxed-spaces, which the writer

filled up with their handwritten characters. The form was made in A4 sized 75 GSM (g/m^2) paper. For writing, we also provided a 0.7 mm ball-point black-inked pen of the same brand to each writer. The writers put the form on a common writing surface while scribbling. All the handwritten pages were scanned at 300 ppi (pixels/inch) in 256 gray-scale with a flat-bed scanner to produce the digital image. The handwritten characters were cropped by a semi-automatic approach with some manual intervention.

Each writer wrote 11 copies of a character, i.e., each writer contributed to a total of 1463 (= 133×11) characters. Multiple copies of the characters were collected to handle the intra-variation of writing [29]. The database collection was performed in a controlled setup, where all writers were aware of our experimentation. With help from some handwriting experts and educational practitioners, we organized a brief training session about handwriting difficulty to the writers. The writers were broadly explained about writing speed/fluency, legibility [30], pen up/down, looping/coiling, etc. Such training helped the writers for self-assessment, since we asked the writers to provide a self-assessment score (A_s) in a continuous range of [0, 10] for each character. We also collected some more information from a writer, e.g., whether the conjunct is familiar to him/her (answer in "Yes"/"No"), writing habit, i.e., approximately how many hours (s)he usually handwrites per day (answer in hours), educational qualification, Secondary School Certificate (SSC)/10th Board examination passing year, gender, and age. We forwarded all these writer-related information parameters including A_s to handwriting experts. On a character image, at least 30 handwriting experts provided their opinion scores in a continuous range of [0, 10], from which we obtained the expert opinion score A_μ. Each expert put his/her opinion scores independently. Finally, the *difficulty score* (a) of a character was computed from A_s and A_μ, as mentioned earlier in Section 3. Thus, every character image ($\mathcal{I}$) of our database was marked with a difficulty score (a).

Writers of our database were mostly hailed from West Bengal, India. After discussing with some educational practitioners, for experimental analysis, we separated these writers into three groups (G_1, G_2, G_3) based on their SSC/10th Board examination passing year, who contributed to our database. G_1 refers to the group of writers who passed SSC before the year 2010. The writers of G_2 passed the SSC in the year 2010 or after 2010. G_3 contains such writers (school students) who have not appeared in the SSC yet during the data collection. Each group contains 30 writers, which leads a total of 90 (= 30×3) writers in our database. The average, minimum, maximum ages of the group G_1 are 42.8, 26, 68 years, respectively. We denote this as G_1: ($avg = 42.8, min = 26, max = 68$). For G_2 and G_3, these measurements are G_2: ($avg = 20.7, min = 16, max = 26$) and G_3: ($avg = 12.2, min = 8, max = 17$), respectively. Such grouping strategy implicitly leads the categorization of writers into various age groups, educational backgrounds, writing habits, etc. The male-female ratios in G_1, G_2, and G_3 groups are 2 : 1, 1 : 1, and 17 : 13, respectively. Some more statistical measures of the writer groups are mentioned in Section 6.1.

Overall, the dataset contains a total of 131670 ($= 90 \times 133 \times 11$) handwritten Bengali characters with corresponding difficulty scores.

5   Proposed Method

In this section, we discuss our proposed solution. Here, we are given a handwritten character image $\mathcal{I}$ as an input, for which we need to generate a difficulty score $\hat{a}$.

After discussing with handwriting experts, educational practitioners, and occupational therapists, we observed that all the fragments of a character are not equally difficult for a writer to scribble. It also holds for Bengali conjunct characters, where multiple characters conjoin. For example, if a writer strains highly to write ক্ষ but not to write ষ, then all the parts of ক্ষ, which conjoins ক and ষ, may not be equally difficult for that writer. This strategy is followed here to find struggling fragments that contribute more to the handwriting difficulty. We compose this as a decision-making task where an agent interacts with a visual environment, viz., handwritten character image to locate the struggling fragments. At every time step, the agent partially observes the input image $\mathcal{I}$ and decides where to focus in the subsequent time step. This problem can be built as a POMDP (Partially Observable Markov Decision Process) [31], where the agent stochastically makes a decision in discrete-time without observing the whole environment [18]. In this work, a reinforcement learning-based agent is employed, which takes action on the environment to learn a policy by maximizing the reward [18]. The major components of POMDP are a set of *states* of the present environment, a set of *actions* to attain the goal, and the *reward* function to optimize decision strategy.

The input character image $\mathcal{I}$ is not resized to avoid any possible character-stroke distortion, which may impede the handwriting difficulty analysis. Instead, we fix the size of the fragments extracted from a character image. The agent partially perceives $\mathcal{I}$ and extracts fragment around a

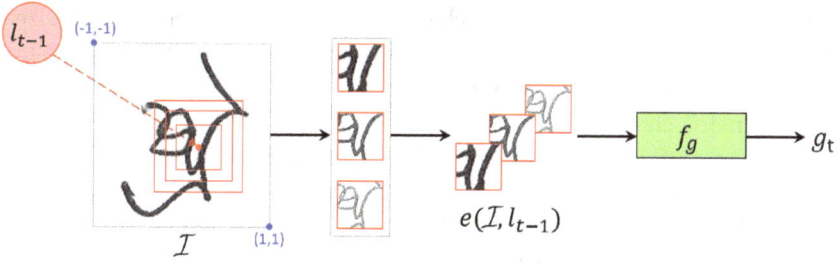

Fig. 1   Retinal transformation

location l. We here adopt the idea of the human visual system regarding foveal and peripheral vision to capture the fine and coarse information [32]. The agent concentrates on the near surroundings of l with a higher resolution, and gradually with a lower resolution far from l [33]. To encode the fragment region, we perform a retinal transformation $e(\mathcal{I}, l)$ on character-image $\mathcal{I}$, which extracts $k > 1$ number of square patches of various resolution by centering location l. The retinal transformation is pictorially shown in Fig. 1. The 1st patch is of size $w \times w$, 2nd patch is of size $(w+d) \times (w+d)$, and so on, the kth patch is of size $(w + (k-1).d) \times (w + (k-1).d)$, where d is the additional width of the consecutive patches. All the k patches are resized to $w \times w$, and thus creates k number of channels each having $w \times w$ sized patch-image. Resizing a larger size patch to a smaller size creates a lower resolution section. In our task, empirically, we decide $k = 3, w = 32$, and $d = 16$. Therefore, the retinal transformation produces a $32 \times 32 \times 3$ sized fragment. The location l is encoded with real-valued co-ordinate (x, y), where $-1 \leq x, y \leq 1$. Here, the centre, top-left and bottom-right co-ordinates of $\mathcal{I}$ are (0, 0), (-1, -1) and (1, 1), respectively. In our task, we randomly initialize the location l on a pixel residing on the character stroke.

Now, $32 \times 32 \times 3$ sized input is fed to f_g, which produces a 512-dimensional feature vector g. This f_g is a CNN [20] containing some convolutional and pooling layers added sequentially, as follows.
Conv (32×32 @ 32) $\Rightarrow$ *MP* (16×16 @ 32) $\Rightarrow$ *Conv* (16×16 @ 64) $\Rightarrow$ *Conv* (16×16 @ 128) $\Rightarrow$ *MP* (8×8 @ 128) $\Rightarrow$ *Conv* (8×8 @ 256) $\Rightarrow$ *MP* (4×4 @ 256) $\Rightarrow$ *Conv* (4×4 @ 512) $\Rightarrow$ *GAP* (1×1 @ 512)
where, "*Conv*" represents the convolutional layer, "*MP*" and "*GAP*" denote max-pooling and global average pooling layers, respectively [20]. The numeric values, in the format of $(m \times m$ @ $c)$, denote the feature map

size $m \times m$ and the number of channels c for a layer. Here, 3×3 sized kernel is employed during all the convolutions, whereas for max-pooling operation, the kernel is of size 2×2. For global average pooling, the 4×4 sized kernel is chosen. For all the convolutions, *swish* activation function is employed [34]. $swish(\mathcal{X}) = \mathcal{X}.\sigma(\mathcal{X})$; where, $\sigma(\mathcal{X}) = (1 + e^{-\mathcal{X}})^{-1}$ is the *sigmoid* activation function. We choose *swish* since it works better than ReLU (Rectified Linear Unit) on the ImageNet database [35]. In our task also, it worked better than ReLU. For adjusting padding and stride in all convolutional layers, we use "same padding" [36]. After the GAP layer, we obtain a 512-dimensional feature vector g by flattening.

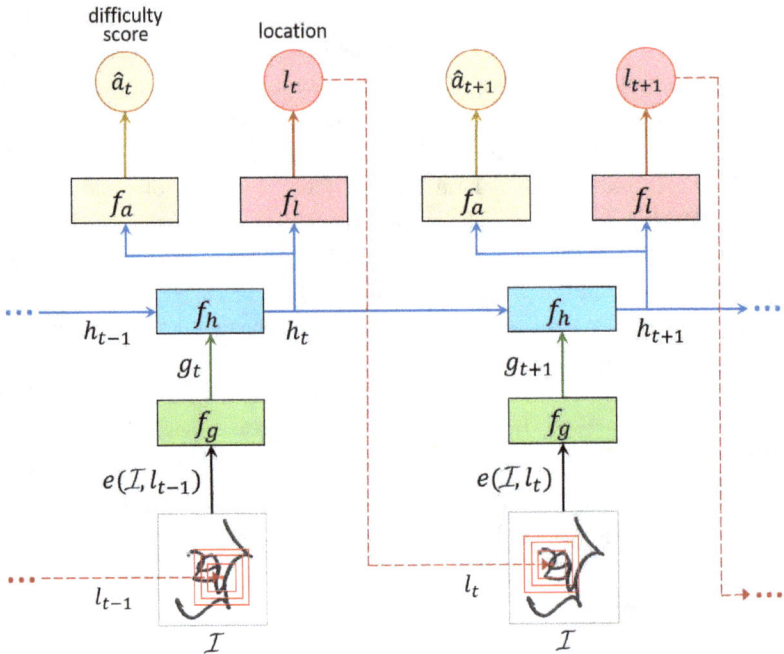

Fig. 2   Proposed architecture

In Fig. 2, we present the workflow of our proposed architecture. At timestep t, f_g produces g_t. This feature vector g_t is fed to our core network f_h which is an RNN [21]. The reason for choosing RNN is to memorize the previously explored fragment information. We use GRU (Gated Recurrent Unit) as a basic RNN unit due to its simplicity and similar performance compared to LSTM (Long Short-Term Memory) on our task [21, 37]. The core network f_h contains 256 GRU units. Here, the agent keeps an internal state which encodes the knowledge extracted from the past locations, and plays a crucial role in deciding the action and finding the next location. The internal/hidden state h_t at time-step t of the RNN is updated over time by the core network f_h. The external input of f_h is g_t. The present hidden state h_t is a function of the previous state h_{t-1} and g_t, which can be written using GRU gates, as follows.

$$h_t = f_h(h_{t-1}, \; g_t) = \Gamma_u * c_t + (1 - \Gamma_u) * h_{t-1} \tag{1.1}$$

where, $\Gamma_u = \sigma(linear(h_{t-1}, \; g_t))$, $c_t = tanh(linear(\Gamma_r * h_{t-1}, \; g_t))$, and $\Gamma_r = \sigma(linear(h_{t-1}, \; g_t))$.

Γ_u and Γ_r represent *update* and *relevant* gates of GRU, respectively [37]. *Sigmoid* (σ) and *tanh* non-linear activation functions are used here [36]. *linear*(v) denotes a linear transformation of vector v.

At this point, h_t is fed to f_a to predict the difficulty score $\hat{a}_t$ from the fragment around location l_{t-1}. $\hat{a}_t = f_a(h_t)$. The f_a follows the concept of deep regression analysis [19], where we perform a linear regression at the rear of a deep architecture. The input layer of f_a, i.e., h_t contains 256 units. After this input layer, we add a fully-connected hidden layer containing 128 nodes with sigmoid activation function. The following output layer of f_a contains one node which is fully connected with the hidden layer with linear activation. The value of this output layer's node is the predicted difficulty score $\hat{a}_t$. Here, we use MAE (Mean Absolute Error) as a loss function due to its robustness to outliers, and backpropagate the loss through f_h and f_g. For learning, minibatch gradient descent with momentum is employed [36]. The gradient of MAE remains the same throughout; therefore, we change the learning rate dynamically using a decaying factor. At timestep t, MAE_t is calculated as follows, $\text{MAE}_t = \frac{1}{n_b} \sum_{i=1}^{n_b} |a_i - \hat{a}_{i,t}|$, where n_b is the number of samples in a batch, a_i is the actual difficulty score of the ith image sample, and $\hat{a}_{i,t}$ is the predicted difficulty score of the ith sample obtained from the fragment around l_{t-1}. As we mentioned earlier, $0 \leq a_i \leq 1$. We employ MAE_t later in reinforcement learning, for the reward shaping.

The h_t is also embedded into f_l to find the next location l. The location policy is defined by a 2-component Gaussian with a fixed variance [33]. We obtain the mean of the location policy at time t from f_l, which is defined as $f_l(h_t) = linear(h_t)$. The f_l is trained using reinforcement learning [18] to find the next location to pay attention to.

In reinforcement learning, the agent interacts with state s of the environment and takes action Ω to get the reward r. At timestep t, the state

s_t takes the fragment centering at location l_{t-1} and summarized into h_t of RNN. The action Ω_t at t is basically the location-action l_t chosen stochastically from a distribution that is θ_l-parameterized by $f_l(h_t)$. We shape the reward r_t at t internally in terms of MAE_t, as follows:

$$r_t = \begin{cases} 1 - \text{MAE}_t & \text{, if } 0 \le \text{MAE}_t < 1 \\ 0 & \text{, otherwise.} \end{cases} \qquad (1.2)$$

Here, $0 \le r_t \le 1$. The highest reward (i.e., $r_t = 1$) is obtained, when $\text{MAE}_t = 0$. For $\text{MAE}_t \ge 1$, no reward is given. It may be noted that the reward upsurges with decreasing MAE value.

The reinforcement learning-based agent learns a θ-parameterized stochastic policy $\pi_\theta(l_t|s_{1:t})$ at every t, which maps the past trajectory of the environmental interactions $s_{1:t}$ to the location-action distribution l_t. For our task, the policy π_θ is defined by early mentioned RNN, and s_t is summarized into h_t. The parameter $\theta = \{\theta_g, \theta_h\}$ is acquired from the parameters θ_g and θ_h of f_g and f_h, respectively. The agent learns to find an optimal policy π^* that maximizes the expected sum of discounted rewards. The cost function J is defined as follows.

$$J(\theta) = \mathbb{E}_{\rho(s_{1:T};\theta)}\left[\sum_{t=1}^{T}\gamma^{t-1}r_t\right] = \mathbb{E}_{\rho(s_{1:T};\theta)}[R] \qquad (1.3)$$

where, ρ is the transition probability from a state to another, which depends on π_θ [18]. T is the total time-step count in an episode and γ is a discounted factor.

Now, to obtain π^*, the optimal parameter is decided by $\theta^* = \arg\max_\theta J(\theta)$. Here, we employ gradient ascent by borrowing the tactics from reinforcement learning literature [38], as follows.

$$\begin{aligned} \nabla_\theta J(\theta) &= \sum_{t=1}^{T}\mathbb{E}_{\rho(s_{1:T};\,\theta)}\left[R\,\nabla_\theta\log\pi_\theta\left(l_t|s_t\right)\right] \\ &\approx \frac{1}{N}\sum_{n=1}^{N}\sum_{t=1}^{T}R^{(n)}\nabla_\theta\log\pi_\theta\left(l_t^{(n)}|s_t^{(n)}\right) \end{aligned} \qquad (1.4)$$

where, trajectories $s(n)$'s are generated by executing the agent on policy π_θ for $n = 1, 2, \ldots, N$ episodes. The $\nabla_\theta\log\pi_\theta(l_t|s_t)$ portion is calculated from the gradient of RNN with standard backpropagation [31].

To avoid the high variance problem of the gradient estimator, variance reduction is performed here [39]. We employ variance reduction with baseline b that comprehends whether a reward is better than the expected, as follows.

$$\nabla_\theta J(\theta) \approx \frac{1}{N} \sum_{n=1}^{N} \sum_{t=1}^{T} \left(R_t^{(n)} - b_t \right) \nabla_\theta \log \pi_\theta \left(l_t^{(n)} | s_t^{(n)} \right) \qquad (1.5)$$

where, $R_t = Q^{\pi_\theta}(s_t, l_t) = \mathbb{E}\left[\sum_{t \geq 1} \gamma^{t-1} r_t | s_t, l_t, \pi_\theta \right]$ and $b_t = V^{\pi_\theta}(s_t) = \mathbb{E}\left[\sum_{t \geq 1} \gamma^{t-1} r_t | s_t, \pi_\theta \right]$, are called as *Q-value function* and *value function*, respectively [18, 39]. It can be observed that Q^{π_θ} depends on l_t, whereas V^{π_θ} does not. The learning of baseline is performed by reducing the squared error between Q^{π_θ} and V^{π_θ}.

Although the idea of discovering the next location is inspired by [33], our architecture of Fig. 2 is quite new. The major differences between our architecture and that of [33] are as follows. In [33], f_g is a shallow neural network, whereas ours is a deep architecture, i.e., a CNN. In f_h, we use GRU as the RNN unit, while [33] employs LSTM. Our f_a is a regression network that generates a difficulty score, whereas [33] uses a softmax layer to execute their classification task. In [33], a global environmental reward is used. For our task, we have proposed an internal reward shaping scheme, which is a new contribution of this chapter.

From our architecture of Fig. 2, we generate k_T number of fragments. Therefore, the total time-step count (T) in an episode is equal to k_T. Empirically, the number of episodes (N) is fixed as 200.

6 Experiments and Discussions

In this section, initially, we discuss the employed dataset and some statistical measures of the writer groups. The data distribution among training, validation, and test set is also mentioned. Then we present the experimental results for handwriting difficulty analysis followed by a comparative analysis with some state-of-the-art approaches.

6.1 *Database employed*

As we mentioned earlier in Section 4, we generated a database of offline Bengali handwritten conjunct characters with corresponding difficulty scores. Our database comprises 3 sets G_1, G_2, and G_3, each of which contains handwritten samples from 30 writers. Every writer wrote 11 copies of 133 distinct characters, i.e., a total of 1463 $(= 133 \times 11)$ characters. Therefore, each set of G_1, G_2, and G_3 contains 43890 $(= 1463 \times 30)$ characters.

Table 1   Statistical measures of writer groups

Group		G_1	G_2	G_3
# writer		30	30	30
male : female		2 : 1	1 : 1	17 : 13
SSC passing year		< 2010	$\geq$ 2010	not appeared
Age (year)	avg	42.8	20.7	12.2
	min	26	16	8
	max	68	26	17
	sd	12.548	3.217	2.524
Writing habit (hours/day)	avg	0.965	3.196	4.6
	min	0.1	1	3
	max	3	6	6.5
	sd	0.723	1.166	0.894
Character familiarity (cf)	avg	0.944	0.914	0.856
	min	0.849	0.796	0.759
	max	1	1	0.962
	sd	0.041	0.057	0.062

avg: average, min: minimum
max: maximum, sd: standard deviation

We obtained some statistical measures, e.g., average (avg), minimum (min), maximum (max), and standard deviation (sd) of the writer groups, concerning writing habit, character familiarity, and age. *Writing habit* is the approximate count of hours per day an individual writes by hand. A writer marked "*Yes*"/"*No*" for every character of the database, when asked whether the conjunct was familiar to him/her. The *character familiarity* of a writer is defined as a ratio of the count of "*Yes*"-marked/familiar characters and the total number of distinct characters in a database.

$$character\ familiarity\ (cf) = \frac{\#familiar\_characters}{\#distinct\_characters\ in\ a\ database} \quad (1.6)$$

where, $\#\ distinct\_characters = 133$ in our database, and $0 \leq cf \leq 1$. From Table 1, we can observe the statistical measures of the writer groups of our database. Based on the average age, $G_1 > G_2 > G_3$. Overall, the younger people write more than the older people per day, i.e., $G_3 > G_2 > G_1$ based on the average writing habit. Based on the average character familiarity, $G_1 > G_2 > G_3$. We observe that a few writers of G_1 and G_2 are familiar with all the characters of our database; for them $cf = 1$. However, their handwriting difficulty scores for some characters were not zero, which suggests that although some writers are familiar with a certain character, they find it difficult to write.

Table 2   Performance of difficulty score prediction

Dataset	MAE	Dataset	MAE
G_1	0.0381	$G_1 + G_2$	0.0392
G_2	0.0326	$G_1 + G_3$	0.0463
G_3	0.0408	$G_2 + G_3$	0.0475
		$G_1 + G_2 + G_3$	0.0498

6.2 Handwriting difficulty analysis

In this subsection, we present the experimental results to see the efficacy of our system. The performance of our system to analyze the handwriting difficulty is evaluated with respect to predicting the difficulty score. Here, we employed MAE (Mean Absolute Error) which is a standard performance measure for the prediction problem. MAE is the arithmetic mean of the absolute differences between actual and predicted difficulty scores of the characters. As a matter of fact, the lower the value of MAE, the better the model is.

As we previously mentioned, our dataset is divided into 3 sets $G_1, G_2,$ and G_3, each of which contains 43890 characters. We performed multiple experiments on the combinations of $G_1, G_2,$ and G_3, as shown in Table 2. For the experiment, the training, validation, and test sets were randomly divided in a ratio of $2 : 1 : 1$.

During the training of our model, some hyperparameters were tuned and fixed. We chose the initial learning rate $= 0.01$, learning rate decay $= 10^{-5}$, momentum $= 0.9$, weight decay $= 10^{-4}$, epoch $= 1000$, and minibatch size $= 128$. We also empirically set $k_T = 20$. In Table 2, we present the system performance of difficulty score prediction in terms of MAE on the test set. Our system achieved 0.0381, 0.0326, 0.0408 of MAE on test sets of G_1, G_2, G_3, respectively. Our system performed the best on G_2. We obtained 0.0498 of MAE on the test set of the entire dataset, i.e., $G_1 + G_2 + G_3$.

6.3 Comparison

To the best of our knowledge, our work is the earliest attempt of its kind. We did not find any related published work from other research groups to make some direct comparisons.

However, for comparative analysis, we extracted the deep feature from a handwritten character image using the front part, i.e., up to the global average pool layer of some major state-of-the-art deep architectures [40–44].

Table 3   Comparison of handwriting difficulty analysis

Dataset	MAE			
Method	G_1	G_2	G_3	$G_1+G_2+G_3$
GoogLeNet [40]	0.0886	0.0851	0.0894	0.0896
ResNet-101 [41]	0.0713	0.0696	0.0745	0.0752
DenseNet-201 [42]	0.0574	0.0542	0.0580	0.0613
Inception-ResNet-v2 [43]	0.0528	0.0507	0.0533	0.0587
EfficientNet-B3 [44]	0.0475	0.0453	0.0497	0.0565
Proposed	0.0381	0.0326	0.0408	0.0498
	Improvement (%)			
$\mathfrak{T}^{\text{proposed}}_{\text{EfficientNet-B3}}$	19.7895	28.0353	17.9074	11.8584

Then the extracted deep feature vector was fed to a regression layer [19] to predict the difficulty score. Our preliminary version of [16] in this research direction used a basic form of GoogLeNet [40]. Therefore, here we compare only with [40] that performed quite similar to [16]. All these competing methods were independent of the reinforcement learning strategy.

In Table 3, we present the comparative study on multiple writer groups of our database. The training, validation, test set distribution remained the same as before (refer to Section 6.2) for all these experiments here. It is observed that our deep reinforcement learning-based method performed better than major state-of-the-art methods concerning difficulty score prediction in terms of MAE. The last row of Table 3 shows the improvement of our model over EfficientNet-B3 [44] that performed the second-best among the methods of this table. The improvement ($\dot{\mathfrak{T}}^{M_1}_{M_2}$) of method-$M_1$ over method-M_2 is calculated as follows.

$$\dot{\mathfrak{T}}^{M_1}_{M_2} = \frac{\text{MAE}_{M_2} - \text{MAE}_{M_1}}{\text{MAE}_{M_2}} \times 100\% \tag{1.7}$$

where, MAE_{M_1} and MAE_{M_2} are the MAE values obtained from methods M_1 and M_2, respectively.

We present $\dot{\mathfrak{T}}^{\text{proposed}}_{\text{EfficientNet-B3}}$ for every writer group in the last row of Table 3. As evident from this table, our proposed method is on average 19.39% better than EfficientNet-B3. This attests to the effectiveness of the reinforcement learning strategy in our model.

7   Conclusion

This section concludes our chapter, where we summarize our current work, provide some direction for our future work, and suggest some new frontiers for future exploration.

7.1 Summary

In this chapter, we have investigated on handwriting difficulty. To inspect difficulties in writing a character, we scrutinized several fragments of the character strokes. We proposed a deep reinforcement learning-based model to detect the fragments that contributed to understanding the handwriting difficulty. Our model employed deep feature-based regression analysis to predict a difficulty measure for a character written by an individual. It also used a recurrent neural network to remember the explored stroke fragments.

For the experimental analysis, we generated a database of Bengali offline handwritten conjunct characters with corresponding difficulty scores. Overall, on the test set of our database, we obtained 0.0498 MAE for predicting the difficulty score, which was quite encouraging. Our model performed better than some deep feature extractor [40–44] with regression analysis.

7.2 Future work

Our research is the earliest attempt that undertakes the handwriting difficulty investigation as a machine learning and pattern analysis problem. In this current chapter, we work on isolated conjuncts. In the future, we will extend this work to examine the conjunct present in a handwritten word, which may be cursive [26]. We will also endeavor to generate more data synthetically, so that our model can learn various facets of challenging handwritten strokes/patterns. Moreover, we intend to test our method on some script other than Bengali to evaluate the system-robustness.

7.3 Frontiers and future scope

In "automated handwriting understanding" [45], the aim is to extract some implicit information from the scribbled strokes. Our research here pushes the frontiers of the automated handwriting stroke understanding, where we predict the difficulty measure from the handwritten character pattern. The applicability of our system is discussed in Section 1. We now discuss some new frontiers relevant to this research.

The handwriting process serves as a link between the human psycho-visual system and bio-motor skill [13,46]. This skill/experience turns out to be important in developing a literate brain [47]. Sometimes, the difficulty in handwriting is observed due to the plausible disturbance of the above-mentioned link. Therefore, the study on handwriting difficulty may open up a new research direction to analyze the psycho-bio-motor process [13].

In this current chapter, we inspect the writing-difficulty on offline handwriting, where a reinforcement learning strategy is adopted to mimic the foveal and peripheral vision in order to gaze the handwriting-image fragments [32]. The handwriting difficulty can be analyzed further with the aid from some electronic devices such as human-eye gaze tracker [48], EEG (Electroencephalography) headset [49], etc.

Our offline handwriting difficulty analyzer can be extended for the online handwriting, where some additional information, e.g., pen pressure, writing direction, velocity, etc., can be obtained from a digital tablet/pen [50]. Sometimes, online handwriting introduces auxiliary difficulties in writing. For example, writing on the smoother surface of a tablet computer becomes challenging for some writers [51]. In a graphic tablet (e.g., Wacom Intuos), handwriting on the decoupled writing surface while viewing the computer screen may create writing-difficulty.

During the data acquisition for the current work, a writer sees a printed version of the character and writes it in his/her own style. This process is termed as "see handwriting" [47], where the writer copies what (s)he sees. We plan to acquire data by "dictation handwriting", where the writer hears from an instructor what to write. During such dictation writing, the writer needs to recollect the character symbols/patterns. Therefore, such writing may introduce some new challenges, where more instances concerning character amnesia are expected [4].

In conclusion, this new problem concerning handwriting difficulty analysis will encourage activities of the DAR community [22].

Acknowledgments

The authors heartily thank all the writers, handwriting experts, and educational practitioners for their contribution to the dataset generation.

References

[1] M. Longcamp, M. T. Z. Poudou, J. L. Velay, "The Influence of Writing Practice on Letter Recognition in Preschool Children: A Comparison Between Handwriting and Typing", Acta Psychologica, vol. 119, pp. 67-79, 2005.

[2] D. Karatzas, S. Robles Mestre, J. Mas, F. Nourbakhsh, P. P. Roy , "ICDAR 2011 Robust Reading Competition - Challenge 1: Reading Text in Born-Digital Images (Web and Email)", ICDAR, pp. 1485-1490, 2011.

[3] H. Ohsaki, K. Tsuda, "Workflow in a Paperless Office", US Patent No.: US6985938B2, Jan. 2006.

[4] C. Hilburger, "Character Amnesia: An Overview", Sino-Platonic Papers, no. 264, pp. 51-70, Dec. 2016.

[5] T. G. Gunning, "Assessing and Correcting Reading and Writing Difficulties", 6th edition, ISBN-13: 9780134516622, Pearson, 2018.

[6] V. W. Berninger, D. T. Mizokawa, R. Bragg, "Scientific Practitioner: Theory-based Diagnosis and Remediation of Writing Disabilities", Journal of School Psychology, vol. 29, no. 1, pp. 57-79, Spring 1991.

[7] D. Ghosh, T. Dube, A. P. Shivaprasad, "Script Recognition - A Review", IEEE Trans. on PAMI, vol. 32, no. 12, pp. 2142-2161, 2010.

[8] P. T. Daniels, W. Bright, "The World's Writing Systems", Oxford University Press, 1996.

[9] R. S. Fu, "Incurable Character Amnesia: The Unavoidable Trend toward the Romanization of Traditional Chinese Handwriting", Sino-Platonic Papers, no. 224, pp. 26-39, Dec. 2012.

[10] D. S. Mather, "Dyslexia and Dysgraphia: More than Written Language Difficulties in Common", Journal of Learning Disabilities, vol. 36, no. 4, pp. 307-317, 2003.

[11] J. Walton, "Handwriting Changes due to Aging and Parkinson's Syndrome", Forensic Sci. Int., vol. 88, no. 3, pp. 197-214, Aug. 1997.

[12] O. Afonso, C. J. Alvarez, C. Martínez, F. Cuetos, "Writing Difficulties in Alzheimer's Disease and Mild Cognitive Impairment", Reading and Writing, vol. 32, pp. 217-233, 2019.

[13] M. A. Bonney, "Understanding and Assessing Handwriting Difficulty: Perspectives from the Literature", Australian Occupational Therapy Journal, vol. 39, no. 3, pp. 7-15, 1992.

[14] M. Hoy, M. Egan, K. Feder, "A Systematic Review of Interventions to Improve Handwriting", Canadian Journal of Occupational Therapy, vol. 78, pp. 13-25, 2011.

[15] P. I. Khalid, J. Yunus, R. Adnan, "Extraction of Dynamic Features from Hand Drawn Data for the Identification of Children with Handwriting Difficulty", Research in Developmental Disabilities, vol. 31, no. 1, pp. 256-262, 2010.

[16] C. Adak, B. B. Chaudhuri and M. Blumenstein, "Cognitive Analysis for Reading and Writing of Bengali Conjuncts", IJCNN, pp. 1-7, 2018.

[17] J. Nickell, "Pen, Ink & Evidence: A Study of Writing and Writing Materials for the Penman, Collector, and Document Detective", ISBN-13: 9781584560920, New Castle: Oak Knoll Press, 2003.

[18] R. S. Sutton, A. G. Barto, "Reinforcement Learning: An Introduction", 2nd ed. Cambridge, MA, USA: MIT Press, 2018.

[19] S. Lathuiliere et al., "A Comprehensive Analysis of Deep Regression", IEEE Trans. on PAMI, vol. 42, no. 9, pp. 2065-2081, 2020.

[20] I. Goodfellow, Y. Bengio, A. Courville, "Deep Learning", MIT Press, 2016. Online: deeplearningbook.org

[21] H. Salehinejad, S. Sankar, J. Barfett, E. Colak, S. Valaee, "Recent Advances in Recurrent Neural Networks", arXiv:1801.01078, 2018.

[22] D. Doermann, K. Tombre, "Handbook of Document Image Processing and

Recognition", ISBN: 9780857298591, Springer, London, 2014.

[23] B. B. Chaudhuri, "Learning an Indian Abugida Script: Bangla", International Graphonomics Society (IGS) Conference, pp. 22-25, 2011.

[24] C. P. Masica, "The Indo-Aryan Languages", Cambridge Language Surveys, Cambridge, 1991.

[25] J. Prinsep, "Inscription in the Old Character on the Rocks of Girnar in Gujerat, and Dhauli in Cuttack", Journal of the Asiatic Society of Bengal, Vol. VII, Pl. XIII, Mar. 1838.

[26] C. Adak et al., "Offline Cursive Bengali Word Recognition using CNNs with a Recurrent Model", ICFHR, pp. 429-434, 2016.

[27] C. Adak, B. B. Chaudhuri, M. Blumenstein, "A Study on Idiosyncratic Handwriting with Impact on Writer Identification", ICFHR, pp. 193-198, 2018.

[28] Q. H. Thu, M. N. Garcia, F. Speranza, P. Corriveau, A. Raake, "Study of Rating Scales for Subjective Quality Assessment of High-Definition Video", IEEE Trans. Broadcasting, vol. 57, no. 1, pp. 1–14, 2011.

[29] C. Adak, B. B. Chaudhuri, C. T. Lin, M. Blumenstein, "Intra-Variable Handwriting Inspection Reinforced with Idiosyncrasy Analysis", IEEE Trans. on Information Forensics and Security, vol. 15, pp. 3567-3579, 2020.

[30] C. Adak, B. B. Chaudhuri, M. Blumenstein, "Legibility and Aesthetic Analysis of Handwriting", ICDAR, pp. 175-182, 2017.

[31] D. Wierstra, A. Foerster, J. Peters, J. Schmidhuber, "Solving Deep Memory POMDPs with Recurrent Policy Gradients", Int. Conf. on Artificial Neural Networks (ICANN), pp 697-706, 2007.

[32] H. Larochelle, G. E. Hinton, "Learning to Combine Foveal Glimpses with a Third-Order Boltzmann Machine", pp. 1243-1251, NIPS, 2010.

[33] V. Mnih, N. Heess, A. Graves, K. Kavukcuoglu, "Recurrent Models of Visual Attention", NIPS, pp. 2204–2212, 2014.

[34] P. Ramachandran, B. Zoph, Q. V. Le, "Swish: A Self-Gated Activation Function", arXiv:1710.05941v1, 2017.

[35] O. Russakovsky et al., "ImageNet Large Scale Visual Recognition Challenge", International J. of Computer Vision, vol. 115, no. 3, pp. 211-252, 2015.

[36] A. Zhang, Z. C. Lipton, M. Li, A. J. Smola, "Dive into Deep Learning", 2020. Online: https://d2l.ai

[37] K. Cho et al., "Learning Phrase Representations using RNN Encoder-Decoder for Statistical Machine Translation", Conf. on Empirical Methods in Natural Language Processing (EMNLP), pp. 1724-1734, 2014.

[38] R. J. Williams, "Simple Statistical Gradient-Following Algorithms for Connectionist Reinforcement Learning", Machine Learning, vol. 8, no. 3-4, pp. 229-256, 1992.

[39] R. S. Sutton et al., "Policy Gradient Methods for Reinforcement Learning with Function Approximation", NIPS, pp. 1057-1063, 1999.

[40] C. Szegedy et al., "Going Deeper with Convolutions", CVPR, pp. 1-9. 2015.

[41] K. He, X. Zhang, S. Ren, J. Sun, "Deep Residual Learning for Image Recognition", CVPR, pp. 770-778, 2016.

[42] G. Huang, Z. Liu, L. V. D. Maaten, K. Q. Weinberger, "Densely Connected Convolutional Networks". CVPR, pp. 2261-2269, 2017.

[43] C. Szegedy, S. Ioffe, V. Vanhoucke, A. A. Alemi, "Inception-v4, Inception-ResNet and the Impact of Residual Connections on Learning", AAAI, pp. 4278–4284, 2017.

[44] M. Tan, Q. V. Le, "EfficientNet: Rethinking Model Scaling for Convolutional Neural Networks", ICML, pp. 6105-6114, 2019.

[45] C. Adak, "A Study on Automated Handwriting Understanding", PhD Thesis, University of Technology Sydney, Australia, 2019.

[46] R. A. Schmidt, "A Schema Theory of Discrete Motor Skill Learning", Psychological Review, vol. 82, no. 4, pp. 225-260, 1975.

[47] K. H. James, "The Importance of Handwriting Experience on the Development of the Literate Brain", Current Directions in Psychological Science, vol. 26, no. 6, pp. 502-508, 2017.

[48] U. Garain, O. Pandit, O. Augereau, A. Okoso, K. Kise, "Identification of Reader Specific Difficult Words by Analyzing Eye Gaze and Document Content", ICDAR, pp. 1346-1351, 2017.

[49] X. Font, A. Delgado, M. F.-Zanuy, "Preliminary Study on the Behavioral Traits Obtained from Signatures and Writing Using Deep Learning Algorithms", in Progresses in Artificial Intelligence and Neural Systems, pp. 199-206, Springer, Singapore, 2020.

[50] E. Anquetil, G. Lorette, "New Advances and New Challenges in On-Line Handwriting Recognition and Electronic Ink Management", in Digital Document Processing, pp. 143-164, Springer, London, 2007.

[51] S. Gerth et al., "Is Handwriting Performance Affected by the Writing Surface? Comparing Preschoolers', Second Graders', and Adults' Writing Performance on a Tablet vs. Paper", Frontiers in Psychology, 7, #1308, 2016.

https://doi.org/10.1142/9789811239014_0007

Chapter 7

Gender Detection from Handwritten Documents Using Transfer Learning Method

Najla AL-Qawasmeh and Ching Y. Suen

Department of Computer Science and Software Engineering,
Concordia University, CENPARMI, Montreal, Canada
n_alqawa@encs.concordia.ca, suen@cse.concordia.ca

Abstract: Offline gender detection from Arabic handwritten documents is a very challenging task because of the high similarity between an individual's writings and the complexity of the Arabic language as well. In this chapter, we propose a new way to detect the writer gender from scanned handwritten documents that mainly based on the concept of transfer-learning. We used a pre-trained knowledge from two convolution neural networks (CNN): GoogleNet, and ResNet, then we applied it on our dataset. We use this two CNN architectures as fixed feature extractors. For the analysis and the classification stage, we used a support vector machine (SVM). The performance of the two CNN architectures concerning accuracy is 80.05% for GoogleNet, 83.32% for ResNet.

Keywords: Handwriting analysis, Transfer-learning, Gender classification, Deep learning

1 Introduction

Gender detection is a biometrical application; hence, Social and physical criteria are considered one of the most important aspects that are used to distinguish between female and male as gender can be detected from different biometric features such as physiological traits (face) or behavioural traits (handwriting) [1] [2]. Despite the high similarity between individual writings, each person has his/her own writing preferences. Many researchers validated the relation between handwriting and gender detection what makes handwriting analysis for gender detection a vital research area in many sectors, including psychology, document analysis, paleography, graphology, and forensic analysis [3] [4] [5]. Computer scientists have also expressed their interest in the handwriting analysis science by developing automatic handwriting analysis systems to detect the writer's gender from scanned images using different machine learning techniques.

The traditional machine learning techniques consist of several complex tasks, including pre-processing, segmentation, feature extraction, which degrade the performance of the systems regarding efficiency and accuracy [6]. The concept of CNN was first introduced in 1980 and 1989 to solve these problems [7] [8]. However, it reached impressive results only in 2012, when Krizhevsky et al. [9] applied a deep convolution network on the ImageNet [10] to classify 1000 different classes.

Deep CNN becomes a very powerful tool for image classification tasks because it automatically extracts the features from raw data, without the need of the traditional ways of machine learning techniques. This makes the feature extraction process easier and faster. On the other hand, CNN does not produce a good performance when the dataset size is small, as the size of the data must be large to get good results [11].

Collecting a large dataset of handwriting is not an easy task, so to solve this issue, the concept of transfer-learning was introduced which mainly depends on using a pre-trained network as a starting point to train a new task. This concept helps in making the process of training using a small number of samples more effective than training a network from scratch [12] [13].

In this proposed work, we used the transfer-learning concept to classify the writer's gender from female and male handwriting images. The main contributions of this paper can be summarized as follow:

- We are proposing a new Arabic handwriting dataset for different classification and recognition tasks.

- We are using the concept of transfer-learning using two different deep-learning architectures.

- We compared the performance of two CNN architectures at the accuracy and precision level.

The rest of the paper is organized as follows: Section 2 explores the related works of some recent automatic handwriting gender detection systems. Section 3 explains the details of the proposed approach. Section 4 describes the experiments and results obtained after applying the proposed method on the target dataset. Section 5 concludes the paper and gives the future directions.

2 Related Works

Interest in gender detection began many years ago, when in 1999, Huber and Headrick [14] presented the correlation between handwriting and several categories, including age, handedness, gender, and ethnicity. However, unfortunately, the old-fashioned way of gender classification suffers from some drawbacks, such as the accuracy of the analysis depends on the analyst's skills, and it is costly and prone to fatigue. These disadvantages created the need to establish an automatic tool, which analyzes the handwriting and detect the writer's gender faster and more precisely with the help of a computer without human intervention [15]. In the following section, we are going to explore some recent automatic Handwriting gender classification systems. Table 1 summaries the gender detection systems reviewed in literature.

Numerous feature types have been extracted from handwritten documents to classify the writer's gender successfully. Some of the classification systems depend on the extraction of geometric features from the handwritings. AlMaadeed and Hussaine [16] proposed an offline automatic handwriting prediction system of age, gender, and nationality based on extracting a set of geometric features from the handwriting images. They combined these features using random forests and kernel discriminative analysis. Their system classification rates were 74.05%, 55.76%, and 53.66% for gender, age, and nationality classification, respectively. Bi et al. [17] presented a new approach for gender detection from handwriting named kernel mutual information (KMI) which focuses on using multi-features to identify the writer's gender. They extracted geometrical features including, slant orientation, curvatures, and transformed features including, Fourier features and WD-LBP features. They used SVM to classify the gender of the writer. Their system achieved an accuracy of 66.3% on ICDAR2013 dataset and accuracy of 66.7% on the registration-document-form (RDF) dataset. Siddiqi et al. [3] proposed an offline prediction system to classify the writer's gender from handwriting images by extracting a set of features such as slant, curvature, texture, and legibility. Artificial neural network (ANN) and support vector machine (SVM) were used for the classification task. They applied their method on two different datasets, QUWI [18] and MSHD [19]. Their approach achieved a classification rate of 68.75 and 67.50 when applying SVM and ANN on QUWI, respectively. Moreover, they produced a classification accuracy of 73.02% and 69.44% when applying SVM and ANN on MSHD dataset, respectively.

Textural features also caught the attention of researchers. Akbari et al. [20] proposed a gender detection system which relies on a global approach that considered the writing images as textures. A probabilistic finite-state automaton was used to generate the feature vectors from a series of wavelet sub-bands. They used SVM and ANN for the classification phase, and they used QUWI and MSHD to evaluate their work. Up to 80% accuracy of classification was achieved in this work. Gattal et al. [15] proposed a gender detection system that uses a combination of different configurations of basic oriented image features to extract the texture features from the handwritten samples. They used an SVM to analyze the extracted features and reveal the writer's gender. Their proposed method achieved classification rates of 71.37%, 76%, and 68% for the ICDAR2013, ICDAR2015, and ICFHR2016 experimental settings, respectively.

Mirza et al. [2] used Gabor-filter to extract the textural features from the handwriting images, to distinguish between male and female writings. The classification task in this work was carried using a feed-forward neural network. They also used QUWI dataset to train and evaluate their method, where they divided the experiments into four tasks. In the first one, they used only Arabic samples for training and testing, while in the second task, they used only English writings for training and testing. However, in the third task, they used Arabic samples for training and English samples for testing. The fourth task was the opposite of the third task, where they used English samples for training and Arabic samples for testing. Their method achieved accuracy rates equal to 70% for task 1, 67% for task 2, 69% for task 3, and 63% for task 4, respectively. Abbas et al. [21] presented a comparative study to evaluate different local-binary methodologies in the classification of gender from handwritings. They proposed a system that presents the images using low- pass filtering features, and then they analyzed these features using the SVM technique. The authors evaluated their method using the QUWI dataset. They achieved 72% of classification accuracy.

Recently, Deep learning becomes a powerful tool for image classification and automatic feature extraction. Illouz et al. [22] proposed a novel deep learning approach for an offline handwritten gender classification system. They used a convolution neural network (CNN) which extracted the features automatically from the handwriting images and classified the writer's gender. The authors created their own dataset, which consists of Hebrew and English handwriting samples of 405 participants. They evaluated their system in two categories: first, Hebrew-English classification, where they trained the system on Hebrew samples and tested it on English samples, and they got a 75.65% of classification accuracy. While in the second category, they trained the system on English samples and tested it on Hebrew samples. The accuracy rate for this category was 58.29%. Morera et al. [23] also used a convolution neural network to detect gender and handedness from offline handwriting. They created their own CNN architecture to classify the handwriting into four multi-classes, including left-handed woman, right-handed woman, left-handed male, and right-handed male. The system achieved an accuracy of 80.72% for gender classification on IAM dataset and accuracy of 68.90% on khatt [24] dataset.

The writer's signatures received the attention of some researchers to classify gender. Maji et al. [25] used handwriting signatures to detect the writer's gender. They extracted several features from the signature's images, including roundness, skewness, kurtosis, mean, standard deviation area, Euler number, the distribution density of black pixels, and the connected components. They used a back-propagation neural network (BPNN) to analyze these features and recognize the writer's gender. Their system achieved an 80.7% accuracy with Euler number feature and a 77% accuracy without the Euler number feature.

Online features have been extracted using special devices such as tablets. Nogueras et al. [1] presented a gender classification schema based on online handwriting. The authors used a digital tablet to extract online features related to the pen trajectory, such as pen-up, pen-down, pen pressure. They achieved a classification accuracy rate of 72.6%.

Navya et al. [26] proposed an offline gender detection system which relies on extracting multi-gradient features from the scanned handwritten documents. They evaluated their method using IAM [27] and QUWI datasets. They achieved 75.6% of classification accuracy. Maken et al. [28] provided a comprehensive review of various techniques used for gender detection systems.

Table 1: Summary of gender detection systems reviewed in literature

	Author	Features extracted	Database Used	Classification rate
1.	AlMaadeed and Hussaine [16]	Geometric features	QUWI	74.06% of gender classification accuracy
2.	Siddiqi et al. [3]	Slant, curvature, legibility	QUWI and MSHD	73.02% and 69.44% of accuracy when they applied SVM and ANN on QUWI and MSHD datasets, resepectively.
3.	Maji et al. [25]	Signature features: roundness, skewness, mean, standard deviation, euler number, black pixels distribution density	They used their own database	80.7% of accuracy with euler feature and 77% of accuracy wiyhout euler feature.
3.	Nogueras et al. [1]	Online features related to pen-trajectory	They collected it using a tablet and stylus	72.6% of accuracy
3.	Mirza et al. [2]	Textural features	QUWI	They got 70% of classification accuracy, when they used only Arabic samples for training and testing
4.	Akbari et al. [20]	textural features	QUWI and MSHD	Up to 80% of classification rate.
5.	Abbas et al. [21]	Low-pass filtering features	QUWI	75.6% of classification rate
6.	Navya et al. [26]	Multi-gradient features	IAM and QUWI datasets	75.6% of classification accuracy
7.	Gattal et al. [15]	Textural features	ICDAR2013, ICDAR2015 and ICFHR2016	They got a classification rates of 71.37%, 76% and 68% when they applied their method on the aforementioned datasets, respectively.
8.	Illouz et al. [22]	Automatically extracted deep features	They used their own dataset	They got 75.65% of classification accuracy, when they trained and tested their method on Hebrew and English samples, respectively.
9.	Morera et al. [23]	Automatic deep features	IAM and Khatt datasets	They got an accuracy rate of 80.72% and 68.90%, when they applied their method on IAM and KHATT datasets, respectively.
9.	Morera et al. [23]	Geometrical features including, slant orientation, curvatures, and transformed features including, Fourier features and WD-LBP features.	registration-document-form (RDF) and ICDAR2013 dataset	Their system achieved an accuracy of 66.3% on ICDAR2013 dataset and accuracy of 66.7% on the registration-document-form (RDF) dataset

3   Methodology

In our proposed method, we used two well-known CNN architectures GoogleNet [29] and ResNet [30] for the detection and the classification of the writer's gender from his/her handwriting. We used the extracted features from each network individually, and then we fed them to an SVM classifier. Figure 1 shows the general structure of our proposed method. The details of each step of our proposed work are given in the following subsections.

3.1 Pre-processing

The pre-processing step in our proposed approach is mainly the resizing of the images to the appropriate size related to each pre-trained network. For ResNet, we resized the images to (223x223x3), and for GoogleNet, we resized the images to (227x227x3).

3.2 Transfer-Learning

Transfer learning is used to speed up and improve training in a new task using the knowledge of a second trained task [13]. We applied the concept of transfer learning using two CNN architectures, as mentioned earlier as fixed feature extractors from the handwritten images. These CNN architectures have been trained over a million images from ImageNet, and they can classify images into 1000 categories.

Training a network deeper makes the training processes very difficult. As the gradients become small, therefore, the weights become unchanged constant, leading to a halt in the network learning process, which affects the network ability to do well. The residual network is a powerful way to solve the vanishing gradient problem. ResNet has residual blocks (skip connections) to jump over layers. It speedup learning by using a fewer number of layers to propagate through [30].

GoogleNet is an inception network. It helps to reduce the computation cost by using 1x1 convolution layer (bottleneck), which is used as a non-linear dimension reduction module [29].

In our proposed work, we replaced the last layers of each pre-trained network (GoogleNet and ResNet) with new fully connected layers adapted to our dataset. The new final layers can extract features that are more relevant to our target task. Then we specify the number of classes in our dataset, which are female and male. We also kept the weights of the transferred layers (early layers) of the pre-trained networks unchanged (frozen) by setting their learning rate to zero, but we increased the learning rate of the new fully connected layers to speed up training and to prevent overfitting to our dataset.

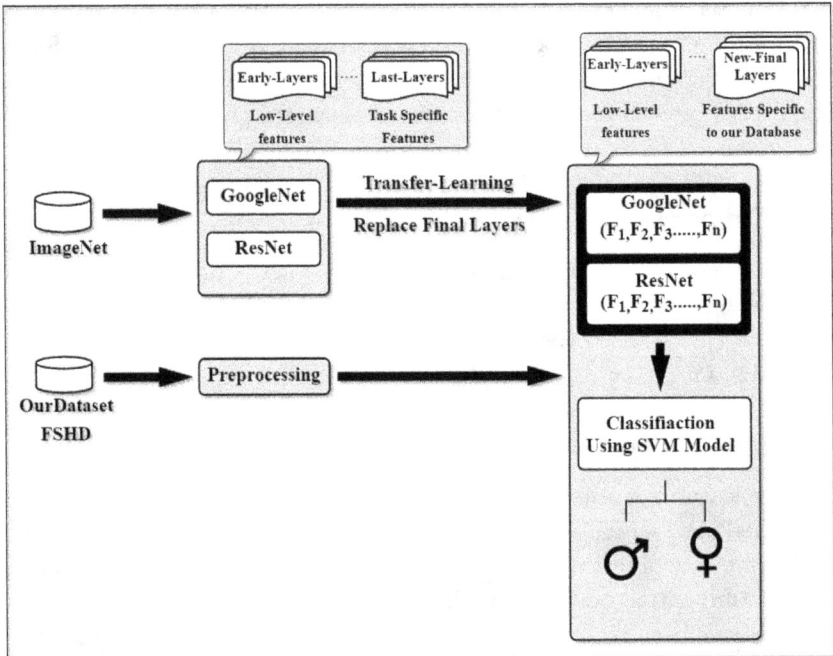

Fig. 1: General structure of the proposed method

4   Experiments and Results

4.1   *Dataset*

As far as we know there is no well-designed and annotated public Arabic handwritten datasets, so we created our own Arabic handwritten dataset to train and test our proposed approach. We called it free-style handwritten samples (FSHS). Hence, datasets should be large enough to cover all the different handwritten styles; we asked 2200 volunteers to write a letter for someone they love. We also asked a portion of the volunteers to copy some paragraphs for future research purposes.

Our dataset in total consists of 2700 handwritten samples which were written by 2200 writers. The writer's ages range from 15 to 75 years old. Men wrote 45% of our dataset, and women wrote the rest. Most of the writer are right-handed, and they are university and school students. Part of the volunteers are employees at some private and governmental companies.

To ensure the convenience of the writers, we did not put any conditions on the type of tool used for writing. However, we provided them with a white sheet of paper to write their letter on it. Volunteers were asked to fill up an information page about their gender, age, handedness, and their work position. We then labeled the dataset concerning the age and the gender of the writer as provided in the information page.

What is significant about our dataset is its size and the number of writers. The variety in age and the large number of writers caused great diversity in the handwriting styles. Table 2 shows a comparison between our dataset and other available datasets. The dataset was mainly written in Arabic, although some writers used the English language, the number of these did not exceed 15 samples. We digitized the handwritten samples using 600 dpi scanner. Our dataset can be used in many research areas related to human interaction, such as handwriting recognition, gender, and age classification. Figure 2 shows some samples of our dataset.

Table 2: Comparison between our dataset and other datasets

	Name of the dataset	Language	Writers	Documents
1.	IFN/ENIT [31]	Arabic	411	2200
2.	AHDB [32]	Arabic	105	315
3.	AWIC2011 [33]	Arabic	54	161
4.	QUWI [18]	Arabic + English	1017	4080
5.	Our dataset(FSHS)	Arabic	2200	2700

4.2 Image classification results

We evaluated our proposed approach using 75% of our dataset for training and 25% for testing. Table 3 shows the evaluation of the SVM classifier with the Gaussian kernel on the extracted features from each pre-trained network (GoogleNet, ResNet) individually in terms of accuracy and precision metrics.

Precision is used to evaluate the cost of false positives predictions as we may lose valuable information if the value is very high. Accuracy is the most common performance metric; it measures the ratio of correct predictions to the total number of input samples to find how often the classifier is correct. These metrics can be computed as follows:

$$Accuracy = \frac{TP + TN}{TP + TN + FP + FN} \tag{7.1}$$

$$Precision = \frac{TP}{TP + FP} \qquad (7.2)$$

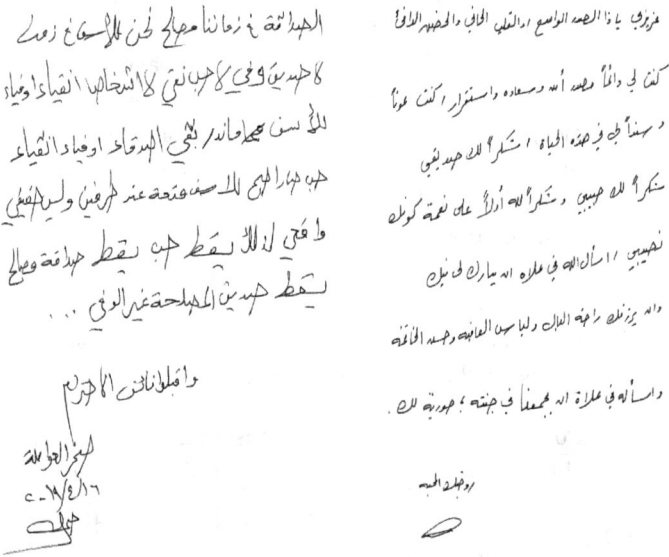

(a) Male   (b) Female

Fig. 2: Samples of the dataset (FSHS)

Table 3: Accuracy and Precision Rates of the Proposed Method

	CNN Architectures	**Accuracy Rates**	**Precision**
1.	**ResNet**	83.32%	82.85%
2.	**GoogleNet**	80.05%	81.18%

5   Conclusions and Future Work

In this research, the concept of transfer learning was used using two CNN architectures ResNet and GoogleNet as fixed feature extractors from handwriting samples. Then SVM was used to analyze the extracted features and classify the writer's gender into female and male. We got a high performance of 83.32% of accuracy when we used the ResNet with its powerful shortcut connections. While the performance of the GoogleNet was equal to

80.05% as the use of a bottleneck layer (1x1 convolutions) helped in the reduction of the computation requirements. For future works, we will enlarge the size of our data samples and add new specific features related to the classification of male and female handwriting to improve the performance of the system.

References

[1] E. Sesa-Nogueras, M. Faundez-Zanuy and J. Roure-Alcobé, Gender classification by means of online uppercase handwriting: A text-dependent allographic approach, *Cognitive Computation* **8**, 1, pp. 15–29 (2016), doi: 10.1007/s12559-015-9332-1.

[2] A. Mirza, M. Moetesum, I. Siddiqi and C. Djeddi, Gender classification from offline handwriting images using textural features, (2016), doi:10.1109/ICFHR.2016.75.

[3] I. Siddiqi, C. Djeddi, A. Raza and L. Souici-Meslati, Automatic analysis of handwriting for gender classification, *Pattern Analysis and Applications* **18** (2014), doi:10.1007/s10044-014-0371-0.

[4] S. Hamid and K. M. Loewenthal, Inferring gender from handwriting in urdu and english, *The Journal of Social Psychology* **136**, 6, pp. 778–782 (1996), doi:10.1080/00224545.1996.9712254, pMID: 9043207.

[5] J. Hartley, Sex differences in handwriting: a comment on spear, *British Educational Research Journal* **17**, 2, pp. 141–145 (1991), doi:10.1080/0141192910170204.

[6] C. Szegedy, S. Ioffe and V. Vanhoucke, Inception-v4, inception-resnet and the impact of residual connections on learning, in *AAAI* (2016).

[7] K. Fukushima and S. Miyake, Neocognitron: A self-organizing neural network model for a mechanism of visual pattern recognition, in S.-i. Amari and M. A. Arbib (eds.), *Competition and Cooperation in Neural Nets.* Springer Berlin Heidelberg, Berlin, Heidelberg, ISBN 978-3-642-46466-9, pp. 267–285 (1982), ISBN 978-3-642-46466-9.

[8] Y. LeCun, B. Boser, J. S. Denker, D. Henderson, R. E. Howard, W. Hubbard and L. D. Jackel, Backpropagation applied to handwritten zip code recognition, *Neural Computation* **1**, 4, pp. 541–551 (1989), doi:10.1162/neco.1989.1.4.541, https://doi.org/10.1162/neco.1989.1.4.541, https://doi.org/10.1162/neco.1989.1.4.541.

[9] A. Krizhevsky, I. Sutskever and G. E. Hinton, Imagenet classification with deep convolutional neural networks, *Commun. ACM* **60**, pp. 84–90 (2012).

[10] J. Deng, W. Dong, R. Socher, L.-J. Li, K. Li and F. F. Li, Imagenet: a large-scale hierarchical image database, pp. 248–255 (2009), doi:10.1109/CVPR.2009.5206848.

[11] S. M. Salaken, A. Khosravi, T. Nguyen and S. Nahavandi, Extreme learning machine based transfer learning algorithms: A survey, *Neurocomputing* **267**, pp. 516–524 (2017), doi:https://doi.org/10.1016/

j.neucom.2017.06.037, `http://www.sciencedirect.com/science/article/pii/S0925231217311657`.

[12] C. Tan, F. Sun, T. Kong, W. Zhang, C. Yang and C. Liu, A survey on deep transfer learning, in V. Kůrková, Y. Manolopoulos, B. Hammer, L. Iliadis and I. Maglogiannis (eds.), *Artificial Neural Networks and Machine Learning – ICANN 2018*. Springer International Publishing, Cham, ISBN 978-3-030-01424-7, pp. 270–279 (2018), ISBN 978-3-030-01424-7.

[13] L. Y. Pratt, J. Mostow, C. A. Kamm and A. A. Kamm, Direct transfer of learned information among neural networks, in *Proceedings of AAAI-91*, pp. 584–589 (1991).

[14] R. Huber and A. Headrick, *Handwriting Identification: Facts and Fundamentals*. CRC Press (1999), ISBN 9781420048773, `https://books.google.ca/books?id=X-2XBySnOQ4C`.

[15] A. Gattal, C. Djeddi, I. Siddiqi and Y. Chibani, Gender classification from offline multi-script handwriting images using oriented basic image features (obifs), *Expert Systems with Applications* **99**, pp. 155 – 167 (2018), doi:https://doi.org/10.1016/j.eswa.2018.01.038, `http://www.sciencedirect.com/science/article/pii/S0957417418300502`.

[16] S. Al Maadeed and A. Hassaine, Automatic prediction of age, gender, and nationality in offline handwriting, *EURASIP Journal on Image and Video Processing* **2014**, 1, p. 10 (2014), doi:10.1186/1687-5281-2014-10, `https://doi.org/10.1186/1687-5281-2014-10`.

[17] N. Bi, C. Y. Suen, N. Nobile and J. Tan, A multi-feature selection approach for gender identification of handwriting based on kernel mutual information, *Pattern Recognition Letters* **121**, pp. 123 – 132 (2019), doi:https://doi.org/10.1016/j.patrec.2018.05.005, `http://www.sciencedirect.com/science/article/pii/S0167865518301715`, graphonomics for e-citizens: e-health, e-society, e-education.

[18] S. Al-ma'adeed, W. Ayouby, A. Hassaïne and M. Jihad, Quwi: An arabic and english handwriting dataset for offline writer identification, (2012), doi:10.1109/ICFHR.2012.256.

[19] C. Djeddi, A. Gattal, L. Souici-Meslati and I. Siddiqi, Lamis-mshd: A multi-script offline handwriting database, (2014), doi:10.1109/ICFHR.2014.23.

[20] Y. Akbari, K. Nouri, J. Sadri, C. Djeddi and I. Siddiqi, Wavelet-based gender detection on off-line handwritten documents using probabilistic finite state automata, *Image and Vision Computing* **59** (2016), doi:10.1016/j.imavis.2016.11.017.

[21] B. Gudla, S. R. Chalamala and S. K. Jami, Local binary patterns for gender classification, in *2015 3rd International Conference on Artificial Intelligence, Modelling and Simulation (AIMS)*, pp. 19–22 (2015), doi:10.1109/AIMS.2015.13.

[22] E. Illouz, E. (Omid) David and N. S. Netanyahu, Handwriting-based gender classification using end-to-end deep neural networks, in V. Kůrková, Y. Manolopoulos, B. Hammer, L. Iliadis and I. Maglogiannis (eds.), *Artificial Neural Networks and Machine Learning – ICANN 2018*. Springer International Publishing, Cham, ISBN 978-3-030-01424-7, pp. 613–621 (2018), ISBN 978-3-030-01424-7.

[23] A. Morera, n. Sánchez, J. Vélez and A. Moreno, Gender and handedness prediction from offline handwriting using convolutional neural networks, *Complexity* **2018**, pp. 1–14 (2018), doi:10.1155/2018/3891624.

[24] S. A. Mahmoud, I. Ahmad, W. G. Al-Khatib, M. Alshayeb, M. T. Parvez, V. Märgner and G. A. Fink, Khatt: An open arabic offline handwritten text database, *Pattern Recognition* **47**, 3, pp. 1096–1112 (2014), doi: https://doi.org/10.1016/j.patcog.2013.08.009, http://www.sciencedirect.com/science/article/pii/S0031320313003300, handwriting Recognition and other PR Applications.

[25] P. Maji, S. Chatterjee, S. Chakraborty, N. Kausar, N. Dey and S. Samanta, Effect of euler number as a feature in gender recognition system from offline handwritten signature using neural networks, (2015).

[26] B. Navya, P. Shivakumara, G. Shwetha, S. Roy, D. Guru, U. Pal and T. Lu, Adaptive multi-gradient kernels for handwriting based gender identification, pp. 392–397 (2018), doi:10.1109/ICFHR-2018.2018.00075.

[27] U.-V. Marti and H. Bunke, The iam-database: an english sentence database for offline handwriting recognition, *International Journal on Document Analysis and Recognition* **5**, 1, pp. 39–46 (2002), doi:10.1007/s100320200071, https://doi.org/10.1007/s100320200071.

[28] P. Maken and A. Gupta, A study on various techniques involved in gender prediction system: A comprehensive review, *Cybernetics and Information Technologies* **19**, pp. 51–73 (2019), doi:10.2478/cait-2019-0015.

[29] C. Szegedy, Wei Liu, Yangqing Jia, P. Sermanet, S. Reed, D. Anguelov, D. Erhan, V. Vanhoucke and A. Rabinovich, Going deeper with convolutions, in *2015 IEEE Conference on Computer Vision and Pattern Recognition (CVPR)*, pp. 1–9 (2015), doi:10.1109/CVPR.2015.7298594.

[30] K. He, X. Zhang, S. Ren and J. Sun, Deep residual learning for image recognition, pp. 770–778 (2016), doi:10.1109/CVPR.2016.90.

[31] M. Pechwitz, S. Snoussi, V. Märgner, N. Ellouze and H. Amiri, Ifn/enit-database of handwritten arabic words, (2002).

[32] J. Ramdan, K. Omar, M. Faidzul and A. Mady, Arabic handwriting data base for text recognition, *Procedia Technology* **11**, pp. 580–584 (2013), doi: https://doi.org/10.1016/j.protcy.2013.12.231, http://www.sciencedirect.com/science/article/pii/S221201731300385X, 4th International Conference on Electrical Engineering and Informatics, ICEEI 2013.

[33] A. Hassaïne, S. Al-ma'adeed, M. Jihad, A. Jaoua and A. Bouridane, The icdar2011 arabic writer identification contest, pp. 1470–1474 (2011), doi: 10.1109/ICDAR.2011.292.

Chapter 8

The Wartegg Test

Graziella Pettinati

Expertise Graziella Pettinati Inc.
info@graziellapettinati.com

Abstract: The Wartegg test is a useful semi-projective drawing test for counseling professionals. The advancement of research in artificial intelligence opens up new perspectives in the analysis and interpretation of the test. An alliance with the research group led by the Italian psychologist Alessandro Crisi could be very profitable.

Keywords: Wartegg test, Artificial intelligence, Drawing analysis, Drawing interpretation

Introduction

The Wartegg's drawing test (WZT, Wartegg Zeichen Test) is a semi-structured drawing test mainly used today in Psychology, Human Resources and Coaching. The test was developed between the 1920's and 30's by the Austro-German psychologist Ehrig Wartegg (1897-1983) [1] [2] [3].

Probably because the test was developed during World War II, it is better known and used in countries closely and indirectly linked to Germany (Latin America, Finland, Italy, Israel and German-speaking countries) and this may explain why the test is practically unknown in English-speaking countries [4] [5].

The evaluation of the drawings of the eight boxes allows drawing a psychological portrait, "head, heart and body".

1. Functional Intelligence
2. Organization of Thought Process
3. Available Energy
4. Social Skills
5. Emotional Functioning, particularly in relationship to the concepts of maturity and affective stability.

1 The Test Itself

The Wartegg Test is a graphic projective test, semi-structured, consisting of a form with 8 squares, numbered from 1 to 8, in two parallel rows, and divided by a black border. In each box there is a specific graphic element; the subject uses this as a starting point and must complete a drawing with meaning in each square.

The sized of the boxes is 4 cm x 4 cm squares in two rows on a black background (Fig. 1) The thickness of the lines is important to visually isolate each of the boxes.

Half of the stimuli are curved based – Boxes 1, 2, 7 and 8 - (Anima) and half are lined based – 3, 4, 5 and 6 (Animus).

The place occupied by each stimulus varies from one box to the other: center, upper left, lower left, upper right, upper center, lower right and upper center.

This is interesting because the theory of the Symbolism of Space was developed at the same time by the German Graphologist, Max Albert Eugene Pulver (1889-1952) [6].

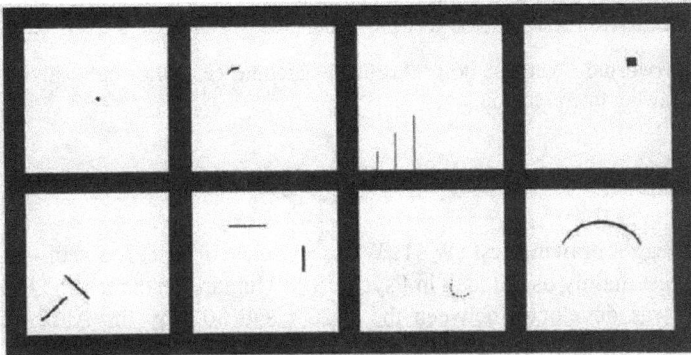

Fig. 1 The 8 boxes of the Wartegg Test

The instructions are easy to follow and quick to administer, most subjects complete it in an average of 5 to 10 minutes. That is why the test is suitable for subjects as young as five (5) years old. The test is also suitable for group administration. The subjects are asked to do the test with a 2HB pencil (no erasers and no rulers are allowed).

The simplicity and neutrality of the stimulus-sign bypasses cultural bias and defensive mechanisms of the subjects.

If the test is not done in front of an evaluator who takes notes (sequence of drawing, hesitation, description of the responses, etc.) while the subject is doing the test, it is important to ask the subject to write down a description on each drawing and the order of his responses.

There are three bases for interpreting the drawings of the Wartegg test [7] [8] [4] [5]:

a) **Relationship** between the stimulus (as an appeal) and the drawing
b) **Content** of the drawing (adequate or inadequate)
c) **Execution** of the drawing

The adequate responses for the stimulus in each box are:

Box 1 Responses where the drawing is centered and/or highly relevant to the drawing, either graphically or thematically.
Box 2 Responses that suggest vitality and movement.
Box 3 Responses that continue the sense of direction indicated by the three existing marks, showing a progression from bottom left to top right, from shorter to taller.
Box 4 Responses that suggest a sense of heaviness, weight, and/or stability.
Box 5 Responses that suggest overcoming an obstacle showing a sense of direction toward the center of the box. (Note that the line starting at the lower left corner is being "blocked" by a second, nearly perpendicular line).
Box 6 Responses where the two marks are joined in some type of synthesis and/or structure.
Box 7 Responses that represent a sense of delicacy.
Box 8 Responses that show a rounding and closure of the curved mark.

The content of the drawings can take various forms:

a) Scribblings (violent or mellow)
b) Abstractions
c) Pictures: animate nature (humans and animals), inanimate nature (landscape), objects, atmosphere and symbolism (cross, flags, etc.)

The execution of the drawings is evaluated with different criteria such as:

a) Good form
b) Balance between curves and lines
c) Filling of space
d) Shading (light, heavy, dark, etc.)
e) Composition (complete or incomplete)
f) Detail (correct, excessive, absent)

The stimulus -sign of each box is derived from gestalt perceptual rules and informs the clinically interpreted meaning as describe in the Table 1.

Table 1. Appearance, Evocative, and Clinical Meaning of the Wartegg Test[a]

Box	Appearance	Evocative	Clinical Meaning
1	A central dot	Centrality, Relevance	Self-Evaluation
2	Away line with curved edges	Vitality, Movement	Female Object Relations
3	Three progressive lines of increasing height	Directionality, Progression Ambition	Available Energy
4	A little black square	Stability, Heaviness	Male Object Relations
5	Two opposite perpendicular lines	Overcoming an Obstacle	Aggressive Energies
6	Two lines at right angles	Synthesis, Structure (Union) Integration	Relation to Reality, Rationality, Planning
7	A dotted semicircle	Delicacy Sociability	Libidinal Energies
8	A convex line	Rounding, Protection, Closure	Social Skills

The two following examples in Fig. 2 and Fig. 3 represent a mother and her son.

---

[a] Crisi, A. (2018) The Crisi Wartegg System (CWS) Manual for Administration, Scoring and Interpretation, p. 190 [8].

Fig. 2 Women 47 years old, right-handed, professional (attorney).
Sequence: 2, 4, 8, 1, 3, 5, 6, 7

Fig. 3 Men 20 years old, right-handed, student – Sequence 3, 2, 7, 6, 8, 5, 4, 1

2 Scoring

Wartegg established five criteria for mainly qualitative evaluation. For each of these criteria, Wartegg has established analysis scoring charts.

1. *Order of drawings*: Boxes 1, 2, 3 and 8 were expected to be drawn first followed by the boxes 4, 5, 6 and 7.

2. *Level of development*: Eight levels of development evaluated by graphic indicators:
    a) First: Scribbling and blackening of the stimulus mark
    b) Second: Failure to integrate the stimulus in the drawing
    c) Third: Integration of the stimulus in the drawing without a particular "meaning"
    d) Forth: Repetition and integration of the stimulus without a clear meaning
    e) Fifth: Increasing of integration of the stimulus
    f) Sixth: Increasing of the integration and more meaningful drawing
    g) Seventh: Drawing related to the subject's experience
    h) Eighth: Drawing showing motives and experience.

3. *Qualitative Profile*: Consistency between the drawings and the expected characteristics associated with each stimulus based on the Gestalt theory. Wartegg qualifies the drawings as "adequate" or "inadequate" according to the way the stimulus is well integrated in coherence with the "nature" of the stimulus. For example, Box 1: organic, stable and natural content is considered adequate while mechanical, dynamic and abstract contents are considered inadequate.

4. *Structure of Boxes*: Structural elements of personality such as perseverative tendencies, personal associations, divergence from the expected order of the drawings, adequate versus inadequate integration for each box were noted.

5. *Characterological Profile*: Wartegg used the theory of personality layers proposed by Lersch (1938) [9] to establish a psychological profile. Each drawing was evaluated using 22 specific criteria. With this evaluation, interpretation can be made regarding moods, feelings, affect, and motivations.

3 Evolution of Scoring and Interpreting the Wartegg Test

Over time, other psychologists in different countries have proposed new ways of assessing and interpreting the Wartegg Test [7] [10].

In 1998, the Italian psychologist Alessandro Crisi, has published the results of his work and research over three decades (more than 30 000 samples) on the Wartegg Test. Dr. Crisi has refined and expanded the scoring and interpretation of the Wartegg Test. The Crisi Wartegg System (CWS) Manual for Administration, Scoring and Interpretation - is available in English since 2018 [8].

The CWS is based on the following two key elements:

1. Two distinct new categories: Evocative Character (EC) and Affective Quality (AQ).

2. Focus on drawings 'Order of Sequence. In 92% of performances, sequence differs from one subject to another. Only 7% of the subjects make the drawings in order (1 to 8). Normal subjects tend to predominantly draw boxes 1, 8, 3 and 6 first (because they concern the ego and its adaptation process), 2 and 4 next (positive compensations); 5 and 7 last (positive compensations).

Evocative Character (EC) points to the capacity of a specific stimulus to recall and facilitate the projection of a particular psychic content.

Affective Quality (AC) is an evaluation exclusively based on the subject's affective connotation of each drawing. It concerns drawings of all contents not only human representation. Allow to assess the emotional disposition of the subjects; the type of affect that characterizes their emotional life; the presence of repression; the degree of harmony that he/she is able to achieve in relationships with the environment; and the presence of depression.

Form quality is based on the criteria of evidence (obviousness).

Affective / Form quality ratio (A/F) – This ratio indicates the presence or absence of an emotional balance.

The sequence of the four pairs (1 & 8), (3 & 6), (2 & 4) and (5 & 7) represents the theoretical model of succession. The order of succession is important. It is a great psychodiagnostic relevance when the subject strays away from this succession. The first and last drawings are more interesting because they indicate more specifically the field in which the subject is more comfortable (first) and the one he prefers to avoid (last).

4 Reliability and Validity of the Crisi Wartegg System

As we know, projective and performance-based personality tests are potentially subjective techniques because their administration and evaluation may be scientifically impacted by factors related to the evaluator (competence, training, relationship to the subject, etc.) and the way the test is being administered. Therefore, it is very important to conduct inter-rater reliability test-retest among scorers [11] [12].

Many studies, both in Italian and English, were conducted on the CWS over more than fifteen years. These studies are described in Chapter 2 of The Crisi Wartegg System (CWS) – Manual for Administration, Scoring and Interpretation [8] [13].

Although new studies are always recommended, especially for subjective psychological tests, the studies conducted on the CWS suggest that there is a convergent validity of the CWS with other established psychological tests such as MMPI-2, the Rorschach Inkblot Test, the AAP, the Guilford-Zimmerman Temperament Survey, and others [13] [14]. The CWS also demonstrated the ability to differentiate between normal and clinical groups, with studies focusing on variety of ages, conditions, and mental health diagnoses [15].

5 Potential Value and Applications

Unfortunately, The Wartegg Test is a well-kept secret for a lot of professionals such as psychologists, psychiatrists, educational and professional counselors, coaches, teachers, social workers, occupational therapists, etc. [8] [4] [10] [16].

The Wartegg Test is a powerful tool that can be very helpful in different working fields:

Clinical: The Wartegg test can be used as well for children, teenagers and adults. The test enables a global description of the personality structure: evaluation of maturity, intellectual functions, potential illness, relationship with parents, relational skills, adjustment to norms, self-confidence, self-esteem, energy level, etc.

Forensic: the Wartegg test can be particularly useful for the analysis of development age subjects in cases of foster care, abuse and violence. In adults, if used with other tests, the Wartegg test contributes to assessing parental abilities and overall psycho diagnostics.

Human Resources: the Wartegg test can be extremely effective as a first screening and allows an evaluation of the candidate in relation to the job description.

Career Counselling (educational and professional): The Wartegg test helps to evaluate the potential, the interests, skills and motivations of a person.

Teaching: The Wartegg test provides a good evaluation of the maturity of the child which is very helpful, especially with young children.

Coaching: The Wartegg test is a relatively quick way to have access to the global personality of the client and enables an objective evaluation if the targeted goals are adequate.

6 Can the Wartegg Test be Automated?

With the advancement of artificial intelligence, the ability to score, analyze and interpret the Wartegg test seems feasible [17]. However, several problems must be overcome.

Collection of all the data: Because the pencil is black and the border of each box is also black, the computer must be able to detect if a stroke has overlapped the border of the box. This element is important for the evaluation of self-control.

Recognition of all the drawings: In order to recognize the drawings in all the boxes it is very important to get a large database of drawing for each stimulus. Moreover, the computer must be able to "detect" the "emotion/affective" of a living drawing (human, animal). For example: Is the dog angry, happy, sitting, moving? Is the landscape peaceful or agitated? When the drawing is not obviously recognizable, is it open to multiple interpretations or unintelligible?

Is the response adequate or inadequate? Once the recognition of the drawing is made, the computer must establish if the response to the stimulus is adequate or inadequate. In other words, is the stimulus well integrated in the drawing or not?

Classification of the drawings: Once the drawing is recognized, it has to be categorized. Crisi has established primary and secondary content categories. These categories are well defined and classified (for example Animal, Abstract, Anatomy, Architecture, Astronomy, Biology, etc.) in the Chapter 4 of his book [8].

Global interpretation: Once all the boxes are evaluated, the analysis must take into account a global evaluation of the eight boxes. This is especially important when a subject makes one drawing with the eight boxes together as seen in (Fig. 5).

The two following examples, in Fig. 4 and Fig. 5, were done by the same person, the same day. The second one, represents a lot of challenges for automated analysis.

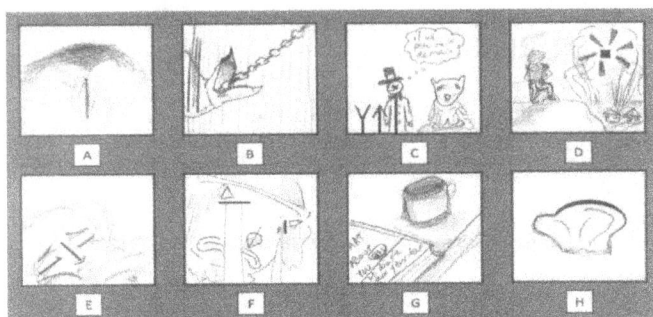

Fig. 4 Man 20 years old, right-handed, student
Sequence: 6, 7, 8, 3, 2, 5, 1, 4

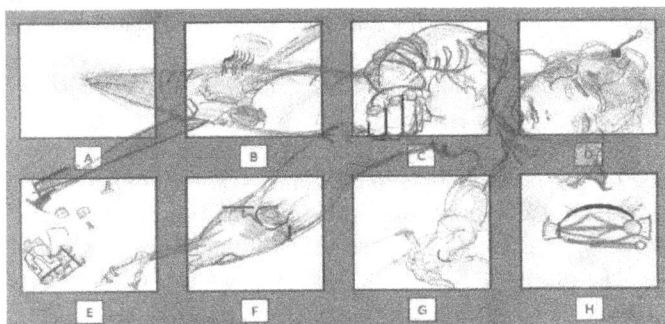

Fig. 5 Man 20 years old, right-handed, student
Sequence 4, 3, 2, 6, 7, 1, 5, 8

Since the CWS has been computerized[b], a collaboration with the Wartegg Italian Institute seems to be the best way to validate the automated system once developed.

The automated Crisi Wartegg System could be a very useful tool for many professionals and would help to get a complete psychological portrait without having to go through a long training process.

---

[b] A demo version of the software can be downloaded at the Instituto Italiano Wartegg web site (http://www.wartegg.com/en/wartegg-test-description/7).

References

[1]   G. M. Kinget, The Drawing-Completion Test; a projective technique for the investigation of personality, New York: Grune & Straton, 1952.

[2]   E. Wartegg, Zeichen der Zeit: Leben und Forschung eines Psychodiagnostikers. [Sign of times: The life and Research of a Ppsychodiagnostician]. An unpublished manuscript from 1977., Mit Ohne Freud. Zur Geschichte des Psychoanalyse in Ostdeutschland [With without Freud: A history of psychoanalysis in Eastearn Germany] ed., H. Bernhart and R. Lockot, Eds., Giessen: Psycosozial-Verlag, 2000, pp. 95–111.

[3]   E. Roivainen, "A Brief History of the Wartegg Drawing Test," *Gestalt Theory*, vol. 31, no. 1, pp. 55-71, 2009.

[4]   U. Avé-Lallemant, Der Wartegg-Zeichentest in der Lebensberatung [The Wartegg Drawing Test in Counselling], München: Reinhardt, 1994.

[5]   M. Gardziella, Wartegg-piirustustesti K. sikirja [The Drawing Wartegg Test. A Handbook], Jyväskylä: Psykologien Kustannus Oy, 1985.

[6]   M. Pulver, Symbolism of Handwriting, L. Herbert, Ed., London: Scriptor Books, 1994.

[7]   P. G. D'Alfonso and C. J. Biedma, Le langage du dessin. Test de Wartegg-Biedma. Version modifiée du test de Wartegg [The Language of Drawing: Wartegg-Biedma Test Modified Version of the Wartegg Test], Paris: Delachaux et Niestlé, Neuchatel - Paris, 1955.

[8]   A. Crisi and J. A. Palm, The Crisi Wartegg System (CWS): Manual for Administration, Scoring, and Interpretation, New-York: Routledge, 2018.

[9]   P. Lersch, Der Aufbau des Charakters. [The Structure of Character], Leipzig: J.A. Barth, 1938.

[10] S. E. Finn, "Using the Crisi Wartegg System in Therapeutic Assessment," *The TA Connection*, vol. 2, no. 1, pp. 12–16, 2014.

[11] R. L. Brennan and D. J. Prediger, "Coefficient Kappa: Some Uses, Misuses, and Alternatives," *Educational and Psychological Measurement*, vol. 41, no. 3, pp. 687–699, 1981.

[12] A. Crisi and F. Dentale, "The Wartegg Drawing Completion Test: Inter-rater agreement and criterion validity of three new scoring categories," *International Journal of Psychology and Psychological Therapy*, vol. 16, no. 1, pp. 85–92, 2016.

[13] J. S. Grønnerød and C. Grønnerød, "The Wartegg Zeichen Test: A literature overview and a meta-analysis or reliability and validity," *Psychology Assessment*, vol. 24, no. 2, pp. 476–489, 2012.

[14] F. Pessotto and R. Primi, "Incremental Validity between the Wartegg and Rorschach tests (R-PAS)," *Paidéia (Ribeirão Preto)*, vol. 31, 2021.

[15] U. Fontana, Una tecnica da non dimenticare. Il WZT validato e allargato per il clinico di oggi [A Method Not to be Forgotten. Validation of the WZT For Today Clinicians], Venice: Editioni Quaderni, 2005.

[16] F. M. Pereira, R. Primi and C. Cobêro, "Validade de testes utilizados em selecao de pessoal Segundo recrudadores [Validity of personal selection tests according to professionals]," *Psychologia: Teoria e Pratica*, vol. 5, no. 2, pp. 83–98, 2003.

[17] L. Liu, G. Pettinati and C. Y. Suen, "Computer-aided Wartegg Drawing Completion Test," in *Second International Conference on Pattern Recognition and Artificial (ICPRAI 2020)*, Montreal, 2020.

Chapter 9

A New Prediction Method for Credit Scoring Based on Sampling Reconstruction of Signal on Graph

Qian Zhang[1], Zhihua Yang[2], Feng Zhou[2] and Lihua Yang[3]

[1]College of Mathematics and Statistics, Shenzhen University

[2]Information Science School, Guangdong University of Finance and Economics

[3]School of Mathematics, Sun Yat-sen University; Guangdong Province Key Laboratory of Computational Science

mcsylh@mail.sysu.edu.cn

Abstract: Credit scoring prediction technique is to assist the lender in making decisions. Accurately predicting user's credit scoring is a challenging issue. In this chapter, a new credit scoring prediction method is proposed, in which the users and their relationships are modeled as a weighted undirected graph, and then the credit scoring prediction problem is reduced to a graph signal reconstruction problem. The proposed method utilizes both the information of the unlabeled data and the location information of the samples in the feature space, and thus achieves an excellent predictive performance. Experimental results on the open UCI German credit dataset are compared with those of seven classical models that the prediction performance of the proposed model is significantly better than that of the reference models. The Friedman test indicates that the experimental results have statistical significance.

Keywords: Credit scoring, Classification, Signal processing on graph

1 Introduction

The credit scoring prediction method is to rate the credit of the loan applicants according to some information, so as to assist the lender in making decisions. An accurate prediction can reduce the cost of lending and the loss of non-performing loans. It is stated that companies could make signif-

icant future savings if an improvement of only a fraction of a percent could
be made in the accuracy of the credit scoring techniques implemented [1].
Therefore, the study of high-precision credit scoring methods has become
a challenging topic in related fields.

Let $V = \{\mathbf{v}_i | \mathbf{v}_i \in \mathcal{R}^m, i = 1, \cdots, n\}$ be the set of users, each user
is described by a m-dimensional feature vector. Suppose each user has a
credit score which is depicted by a real, then the scores of all the users form
the set $\mathcal{Y} = \{y_i | y_i \in \mathcal{R}, i = 1, \cdots, n\}$. For convenience, in the following
we name a score as a label. In our problem, only part of the users's label
is known. We call these users the labeled ones, and others the unlabeled
ones. Let $\mathcal{K} = \{\mathbf{v}_i, \ i = \ell_1, \cdots, \ell_l\} \in V$ be the set of all the labeled users.
The basic question of the credit scoring prediction is to correctly predict
the values of the labels y_i for the unlabeled users, $\mathbf{v}_i \notin \mathcal{K}$.

In the past few decades, many scholars have put forward many credit
scoring methods. They can be divided into two categories: single model and
ensemble model. The single model is mainly based on statistical and machine learning methods. Because the credit score is usually normalized as a
non-negative integer in practical applications, the problem of credit scoring
is essentially a classification problem. The goal of credit scoring method
is to generate classifiers according to the distribution of known good users
and bad users in feature space in order to predict the credit level of unknown users accurately. Therefore, the traditional classification methods
are usually used to design the credit score methods. Typical representatives
include linear discriminant analysis (LDA) [2], logistic regression (LR) [3],
decision trees (DT) [4], naive Bayes (NB) [5], neural network (NN) [6], support vector machine (SVM) [7] and random forest (RF) [5] et al. The idea
of ensemble models is to combine the results of multiple single classifiers,
called a base classifier, to produce a comprehensive score. Therefore, the
selection of base classifiers should be as good and different as possible, that
is, to improve the accuracy and diversity of base classifiers as much as possible [8]. At present, most ensemble models use logical regression, support
vector machine and so on as the base classifiers of ensemble models [3].

From the point of view of machine learning, almost all existing methods
fall into the category of supervised learning. This means that all unknown
samples do not participate in the training of classifiers at all. Recently,
semi-supervised learning techniques have made great progress [9]. Researches have found that unlabeled data, when used in conjunction with
labeled data, can produce considerable improvement in learning accuracy.
This idea motivates us to make use of the information of unknown users to

participate the credit scoring prediction.

To take advantage of the information carried by the unknown users, we model the users as well as their relationships as an undirected and weighted graph, $\mathcal{G} = (\mathcal{V}, \mathbf{W})$, in which $\mathcal{V}$ is a set of vertices, one vertex indicates a user and the weight matrix $\mathbf{W}$ represents the connections or similarity between the vertices. With this model, the problem of predicting credit of unknown users becomes a sampling reconstructive problem of signal on graph. On the other hand, we map a vertex which is represented by a feature vector into a high-dimensional feature space through a feature mapping. According to the manifold assumption [9], the images of all the vertices in the high dimensional feature space do not uniformly fill the entire feature space and are usually distributed on a low dimensional manifold in an implicit and easy-to-classification manner. This means that the signal which is composed of vertex labels is continuous in the feature space, which inspires us to find the reconstruction signal in a reproducing kernel Hilbert space. Finally, the original problem is approximately modeled as a quadratic unconditional optimization problem in a reproducing kernel Hilbert space. The famous Representer Theorem of the Reproducing Kernel Hilbert Space (RKHS) is employed to get the analytical solution. We conduct experiments on the open UCI German credit dataset, and the encouraging results are received.

The rest of the chaper is organized as follows: Section 2 includes the main contributions of this chapter, which discusses the prediction model of signal on graph and its solutions, and gives the new credit scoring model. In Section 3, the details of experiment are introduced and the experimental results are exhibited. Finally, a brief conclusion of the chapter is given in Section 4.

2   The Prediction Method Base on Sampling Reconstruction of Signal on Graph

In this section, we discuss how to use the manifold assumption to model the prediction problem as a sampling reconstruction of signal on graph, as well as its solving strategies.

2.1   *Mathematical Formulation of the Prediction Problem*

Semi-supervised learning is based on three types of model assumptions: namely, the smoothing hypothesis, clustering hypothesis, and manifold hypothesis, to make using unlabeled data to help improve learning performance [9]. The manifold hypothesis assumes that the data is distributed on a manifold structure, and adjacent samples have similar label values. Since the proximity of the samples can be described by graphs, the manifold hypothesis is widely used in semi-supervised learning.

Let $\mathcal{V}$ be the set of all the users and $\mathbf{W}$ be the adjacent matrix, whose entry w_{ij} is a nonnegative real number which expresses the affinity between users $\mathbf{v}_i$ and $\mathbf{v}_j$. Then $\mathcal{G} := (\mathcal{V}, \mathbf{W})$ makes up an undirected weighted graph. According to the notations of graph theory [12, 13], the set $\mathcal{V}$ and the matrix $\mathbf{W}$ are called the vertex set and adjacent matrix of the graph $\mathcal{G}$, respectively. Each entry w_{ij} of the adjacent matrix is called the weight between the vertice $\mathbf{v}_i$ and $\mathbf{v}_j$.

Under the framework of the weight graph $\mathcal{G} := (\mathcal{V}, \mathbf{W})$, the labels of all vertices can be represented by a function

$$\mathbf{s} : \mathcal{V} \mapsto \mathcal{R},$$

where $\mathbf{s}(\mathbf{v}_i)$ indicates the label of vertex $\mathbf{v}_i$. Now we consider such a problem: given the function values on some vertices, we want to find a function such that the function values at these vertices are equal to the given labels. Therefore, once such a function can be found, the function values on the unlabelled vertices can be used as an estimate of their labels. This idea is described as follows:

$$\text{Find } \mathbf{s} : \mathcal{V} \to \mathcal{R}, \quad \text{subject to } \mathbf{s}(\mathbf{v}_j) = y_j, \ \mathbf{v}_j \in \mathcal{K}. \tag{9.1}$$

As we mentioned above, connection or similarity is a measure of the affinity between two vertices. If two vertices possess strong connection then their labels should be close. This means that the function $\mathbf{s}$ should be continuous on the graph. According to the well-known manifold assumption, vertices with large weights should have similar values [10]. This observation induces us to introduce the following continuity measure:

$$\sigma(\mathbf{s}) := \frac{1}{2} \sum_{i,j=1}^{n} w_{ij}((\mathbf{s}(\mathbf{v}_i) - \mathbf{s}(\mathbf{v}_j))^2. \tag{9.2}$$

Let $\mathbf{L}$ be the Laplacian matrix of the graph defined by $\mathbf{L} := \mathbf{D} - \mathbf{W}$, where $\mathbf{D}$ is the degree matrix of the graph. A simple calculation deduces that the smoothness measurement $\sigma(s)$ defined by Eq. (9.2) can be rewritten as [17]:

$$\sigma(s) = \mathbf{s}^T \mathbf{L} \mathbf{s}, \quad \text{for } \mathbf{s} := (s(\mathbf{v}_1), \cdots, s(\mathbf{v}_n))^T \in \mathcal{R}^n, \tag{9.3}$$

where the superscript T stands for the transpose operation of the matrix.

To find the most continuous solution of Eq. (9.1), we consider the following modified problem

$$\min_{\mathbf{s}:\mathcal{V}\to\mathcal{R}} \mathbf{s}^T\mathbf{L}\mathbf{s}, \quad \text{subject to} \quad \mathbf{s}(\mathbf{v}_j) = y_j, \quad \mathbf{v}_j \in \mathcal{K}. \tag{9.4}$$

This optimization problem can be approximately converted to the following unconstrained optimization problem by introducing a penalty item,

$$\min_f r(f) := \sum_{\mathbf{v}_j \in \mathcal{K}} |f(\mathbf{v}_j) - y_j|^2 + \gamma \mathbf{f}^T\mathbf{L}\mathbf{f}, \tag{9.5}$$

where $\mathbf{f} := (f(\mathbf{v}_1), \cdots, f(\mathbf{v}_n))^T$, the penalty term $|f(\mathbf{v}_j) - y_j|^2$ is used to ensure that the value of f at the vertex $\mathbf{v}_j$ is approximately equal to y_j. The regularization term $\mathbf{f}^T\mathbf{L}\mathbf{f}$ makes the solution f be as smooth as possible and $\gamma > 0$ is a regularization parameter which controls the importance of the regularization term and the penalty terms. It is easy to see that the solution of Eq. (9.5) is an approximation to that of Eq. (9.4). The smoothness of the solution depends on the value of the regularization term γ. Here we take the notation f instead of $\mathbf{s}$ in Eq. (9.4) because the solution f of Eq. (9.5) may not satisfy the constraints in Eq. (9.4). Strictly speaking, f is not the solution of Eq. (9.4), it is only an approximate solution.

2.2 *Solution of the Optimization Objection*

In semi-supervised leaning, the unlabelled data is also considered to carry useful information, which forms the basic assumption of unsupervised learning. We assume all the vertices in $\mathcal{V}$ can be mapped into a feature space of higher dimension by a feature mapping Φ. According to the manifold assumption [9], the images of the vertices under Φ will not fill the whole feature space uniformly, they usually distributed on a low-dimensional manifold in an implicit and easy-to-classify way. It means that the function $\mathbf{s}$ is continuous on the feature space. Let

$$\kappa(\mathbf{v}_i, \mathbf{v}_j) := \Phi(\mathbf{v}_i)^T\Phi(\mathbf{v}_j), \quad \forall \mathbf{v}_i, \mathbf{v}_j \in \mathcal{V}.$$

Then $\kappa : \mathcal{V} \times \mathcal{V} \to \mathcal{R}$ is a symmetric and positively definite function and thus defines a reproducing kernel Hilbert space

$$\mathcal{H} := \Big\{ \sum_{i=1}^n a_i \kappa(\cdot, \mathbf{v}_i) | a_1, \cdots, a_n \in \mathcal{R} \Big\},$$

with inner products

$$\Big\langle \sum_{i=1}^n a_i \kappa(\cdot, \mathbf{v}_i), \sum_{j=1}^n b_j \kappa(\cdot, \mathbf{v}_j) \Big\rangle := \sum_{i=1}^n \sum_{j=1}^n a_i b_j \kappa(\mathbf{v}_i, \mathbf{v}_j).$$

According to the reproducing property we have that

$$f(\mathbf{v}) = \langle f, \kappa(\cdot, \mathbf{v}) \rangle, \quad \forall \mathbf{v} \in \mathcal{V}.$$

Thus, by the Cauchy-Schwartz inequality we have

$$|f(\mathbf{v}_i) - f(\mathbf{v}_j)| = |\langle f, \kappa(\cdot, \mathbf{v}_i) - \kappa(\cdot, \mathbf{v}_j) \rangle| \leq \|f\|_{\mathcal{H}} \|\kappa(\cdot, \mathbf{v}_i) - \kappa(\cdot, \mathbf{v}_j)\|_{\mathcal{H}},$$

where $\|\cdot\|_{\mathcal{H}}$ denotes the norm in $\mathcal{H}$. Since

$$\kappa(\mathbf{v}_i, \mathbf{v}_j) = \langle \kappa(\cdot, \mathbf{v}_i), \kappa(\cdot, \mathbf{v}_j) \rangle,$$

$\Phi(\mathbf{v}) := \kappa(\cdot, \mathbf{v})$ can be regarded as a feature mapping. Thus

$$|f(\mathbf{v}_i) - f(\mathbf{v}_j)| \leq \|f\|_{\mathcal{H}} \|\kappa(\cdot, \mathbf{v}_i) - \kappa(\cdot, \mathbf{v}_j)\|_{\mathcal{H}} = \|f\|_{\mathcal{H}} \|\Phi(\mathbf{v}_i) - \Phi(\mathbf{v}_j)\|_{\mathcal{H}}.$$

This shows that the norm $\|f\|_{\mathcal{H}}$ characterizes the continuity of f on $\Phi(\mathbf{v})$.

To guarantee the continuity of the solution of (9.5) in the feature space, we choose a suitable reproducing kernel Hilbert space $\mathcal{H}$, insert the term $\|f\|_{\mathcal{H}}^2$ into the objective function of Problem (9.5) and solve it in $\mathcal{H}$. Thus, Problem (9.5) is modified as

$$\min_{f \in \mathcal{H}} r(f) := \sum_{\mathbf{v}_j \in \mathcal{K}} |f(\mathbf{v}_j) - y_j|^2 + \gamma \mathbf{f}^T \mathbf{L} \mathbf{f} + \lambda \|f\|_{\mathcal{H}}^2, \tag{9.6}$$

where $\lambda > 0$ is a regularization parameter.

In order to solve the above optimization problem, we introduce the famous theorem in RKHS, that is the representer theorem [10]:

Theorem 2.1 (Representer Theorem). *Let $\mathcal{V}$ be a nonempty set and κ a reproducing kernel on $\mathcal{V} \times \mathcal{V}$ with corresponding reproducing kernel Hilbert space $\mathcal{H}$. Given a training sample set including both labeled and unlabeled samples. Without loss of generality, let its first l samples be labeled ones, i.e., $\{(\mathbf{v}_1, y_1), \cdots, (\mathbf{v}_l, y_l)\} \in \mathcal{V} \times \mathcal{R}$. Then for any function $c : (\mathcal{V} \times \mathcal{R})^l \to \mathcal{R} \cup \{\infty\}$ and monotonic increasing function $\Omega : [0, \infty) \to \mathcal{R}$, each minimizer $f \in \mathcal{H}$ of the regularized risk functional*

$$c((y_1, f(\mathbf{v}_1)), \cdots, (y_l, f(\mathbf{v}_l))) + \Omega(\|f\|_{\mathcal{H}}) + \gamma \mathbf{f}^T \mathbf{L} \mathbf{f}, \tag{9.7}$$

admits an expansion [18]

$$f(\mathbf{v}) = \sum_{i=1}^{n} a_i \kappa(\mathbf{v}_i, \mathbf{v}) \tag{9.8}$$

in terms of both labeled and unlabeled samples.

Denoting

$$\mathbf{K} = \begin{bmatrix} \kappa(\mathbf{v}_1, \mathbf{v}_1) & \cdots & \kappa(\mathbf{v}_1, \mathbf{v}_n) \\ \vdots & & \vdots \\ \kappa(\mathbf{v}_n, \mathbf{v}_1) & \cdots & \kappa(\mathbf{v}_n, \mathbf{v}_n) \end{bmatrix} \tag{9.9}$$

as the kernel gram matrix and

$$\mathbf{f} = \begin{bmatrix} f(\mathbf{v}_1) \\ \vdots \\ f(\mathbf{v}_n) \end{bmatrix}, \quad \mathbf{a} = \begin{bmatrix} a_1 \\ \vdots \\ a_n \end{bmatrix}, \tag{9.10}$$

then the solution of Eq. (9.6) can be written as

$$\mathbf{f} = \mathbf{K}\mathbf{a}. \tag{9.11}$$

Without loss of generality, we assume that $\{\mathbf{v}_j\}_{j \in J}$ for $J := \{1, \cdots, l\}$ is the set of the labelled vertices and $\{y_j\}_{j \in J}$ is the corresponding labels set. Let $\mathbf{y}_J := (y_1, \cdots, y_l)^T$ and $\mathbf{K}_J$ the sub-matrix consisting of the first l rows of $\mathbf{K}$. It is easy to see that $\mathbf{K}_J \mathbf{a} - \mathbf{y}_J$ is a l-dimensional vector whose ith element is

$$(\mathbf{K}_J \mathbf{a} - \mathbf{y}_J)_i = \sum_{j=1}^{n} \kappa(\mathbf{v}_i, \mathbf{v}_j) a_j - y_i = f(\mathbf{v}_i) - y_i, \; i \in J.$$

Thus the Euclidean norm of $\mathbf{K}_J \mathbf{a} - \mathbf{y}_J$ in $\mathcal{R}^l$ is

$$\|\mathbf{K}_J \mathbf{a} - \mathbf{y}_J\|_2^2 = \sum_{i \in J} |f(\mathbf{v}_i) - y_i|^2. \tag{9.12}$$

On the other hand, according to [10], the norm of f in $\mathcal{H}$ is

$$\|f\|_{\mathcal{H}}^2 = \sum_{i=1}^{n} \sum_{j=1}^{n} a_i a_j K(\mathbf{v}_i, \mathbf{v}_j) = \mathbf{a}^T \mathbf{K} \mathbf{a}. \tag{9.13}$$

According to Eq. (9.11), Eq. (9.12) and Eq. (9.13), the optimization problem Eq. (9.6) can be reduced to the following equivalent problem

$$\min_{f \in \mathcal{H}} J(f) := \|\mathbf{K}_J \mathbf{a} - \mathbf{y}_J\|_2^2 + \gamma \mathbf{a}^T \mathbf{K}^T \mathbf{L} \mathbf{K} \mathbf{a} + \lambda \mathbf{a}^T \mathbf{K} \mathbf{a}. \tag{9.14}$$

Using $\|\mathbf{K}_J \mathbf{a} - \mathbf{y}_J\|_2^2 = (\mathbf{K}_J \mathbf{a} - \mathbf{y}_J)^T (\mathbf{K}_J \mathbf{a} - \mathbf{y}_J)$, the objective function of Eq. (9.14) can be further simplified in form of

$$J(f) = \mathbf{a}^T \left(\mathbf{K}_J^T \mathbf{K}_J + \gamma \mathbf{K}^T \mathbf{L} \mathbf{K} + \lambda \mathbf{K} \right) \mathbf{a}$$
$$-2(\mathbf{K}_J^T \mathbf{y}_J)^T \mathbf{a} + \|\mathbf{y}_J\|_2^2. \tag{9.15}$$

Since $\mathbf{L}$ is a semi-positive matrix, if $\mathbf{K}$ is positive definite and $\lambda > 0$, then $\mathbf{K}_J^T \mathbf{K}_J + \gamma \mathbf{K}^T \mathbf{L} \mathbf{K} + \lambda \mathbf{K}$ is positive definite. Therefore, $J(f)$ is the convex quadratic function of $\mathbf{a}$ which implies that the solution of Eq. (9.14) exists uniquely and can be expressed analytically as:

$$\mathbf{a} = \left(\mathbf{K}_J^T \mathbf{K}_J + \gamma \mathbf{K}^T \mathbf{L} \mathbf{K} + \lambda \mathbf{K} \right)^{-1} \mathbf{K}_J^T \mathbf{y}_J. \tag{9.16}$$

Once $\mathbf{a}$ has been obtained, we can get the optimal predicted values on all vertices from Eq. (9.11).

2.3 *A New Credit Prediction Algorithm*

In this section, we will discuss how to apply the vertex label prediction method proposed in the previous section to the credit scoring prediction problem.

Based on the discussion above, we first need to model the user and their relationship as a weighted undirected graph. To do so, the most key process is to establish the weight matrix $\mathbf{W}$. In fact, the weight matrix $\mathbf{W}$ plays a crucial role in the proposed method. Its construction has become an important issue in the signal processing on graphs [12]. So far, there are two strategies: The first one is by means of prior domain knowledge [13]; The second one is learning from observed data [14].

The next issue is to choose the kernel function $\kappa : \mathcal{V} \times \mathcal{V} \to \mathcal{R}$, or equivalently matrix $\mathbf{K} = (\kappa(\mathbf{v}_i, \mathbf{v}_j))_{i,j \in \mathcal{V}}$, where $\mathbf{v}_i$ and $\mathbf{v}_j$ are the feature vectors of the users i and j. In this chapter, we choose the most commonly used Gaussian kernel, that is,

$$\kappa(\mathbf{v}_i, \mathbf{v}_j) = \exp(-\|\mathbf{v}_i - \mathbf{v}_j\|^2 / \sigma^2), \tag{9.17}$$

with the constant parameter σ.

Once the graph structure is established, the problem of user credit scoring prediction can be solved by the above method of vertex label prediction on the graph. The strategy of credit scoring prediction is illustrated in Figure 1.

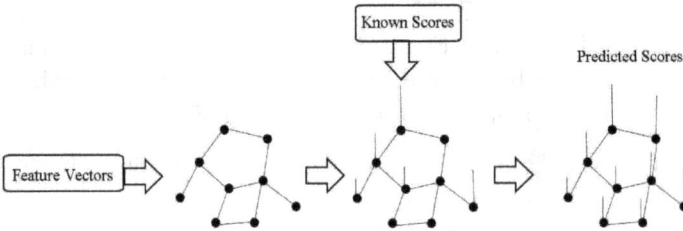

Fig. 1   Diagram of the credit scoring model.

Now, we propose the following prediction algorithm.

Algorithm 2.1. Given matrix $\mathbf{V}_{n \times m}$ which consists of the feature vectors of all users and whose first l rows indicate the known users, $\mathbf{y}_J = (y_1, y_2, \cdots, y_l)^T$ is the vector composed of the labels of the known users, γ and λ are the regularization parameters.

(1) Compute the adjacent matrix $\mathbf{W}$ and Laplacian matrix $\mathbf{L}$ to form an undirected and weighted graph $\mathcal{G} := \{\mathcal{V}, \mathbf{W}\}$, where $\mathcal{V} := \{\mathbf{v}_1, \cdots, \mathbf{v}_n\}$ is the set of vertices.
(2) Compute the kernel matrix $\mathbf{K}$ by Eq. (9.9), and (9.17).
(3) Pick out the first l rows of $\mathbf{K}$ to form the sub-matrix $\mathbf{K}_J$.
(4) Compute $\mathbf{A} := \mathbf{K}_J^T \mathbf{K}_J + \gamma \mathbf{K}^T \mathbf{L} \mathbf{K} + \lambda \mathbf{K}$;
(5) Solve $\mathbf{A}\mathbf{a} = \mathbf{K}_J^T \mathbf{y}_J$ to get $\mathbf{a}$ and the compute $\mathbf{f} = \mathbf{K}\mathbf{a}$.

The last $n - l$ items are the predicted labels of the unknown users.

3 Experimental Results and Analysis

In this section, experimental results are presented to demonstrate the superiority of the proposed method. The experiments are conducted on the famous UCI German Credit dataset which is publicly available at the UCI repository (http://kdd.ics.uci.edu/). This dataset contains 1000 instances, of which 70 percent are good users and 30 percent are bad users. It provides two files: the file "german.data" is the original dataset which includes 20 attributes (7 numerical, 13 categorical), and the file "german.data-numeric" is the numerical version, in which the original attributes have been edited and several indicator variables have been added to form the final 24 numerical attributes to make it suitable for algorithms. In the experiments, we adopt the numerical version.

In the original data, there are large order of magnitude differences among the data of different attributes. In order to eliminate the influence of the order of magnitude difference between the data on the classification, we first preprocess the original data as follows:

$$v_{i,k} = \frac{v_{i,k} - v_{kmax}}{v_{kmax} - v_{kmin}},$$

where $v_{i,k}$ indicates the ith user's the kth attribute, v_{kmax} and v_{kmin} are the maximum and the minimum of the kth attribute of all the users, respectively.

We first divide the entire data into two subsets at a ratio of 80%/20%, the bigger one for training and another for testing. We call this pair of subsets the validation set, and they will be used to find suitable parameters. Then, we randomly and evenly divided the entire data set into five subsets, namely each contains 20% users. These subsets will be used to conduct a 5-folds cross-validation.

The preprocessed data directly is served as the feature vector. A good user is given the label "1" and a bad user "0". The function *gsp_nn_graph()*

of the open tool kit for graph signal process, DRASP [19], is called to build a k-nearest neighbors graph.

For the built graph, there are four parameters that will affect the final prediction performance. We conduct experiments on the validation set to search for appropriate parameters. After a larger number of experiments, we finally select the parameters as follows: $\lambda = 0.5, \gamma = 0.025, \sigma = 1$ *and* $k = 20$. Unless otherwise specified, the following experiments will use these parameters. Figure 2 $\sim$ Figure 5 show some experimental results.

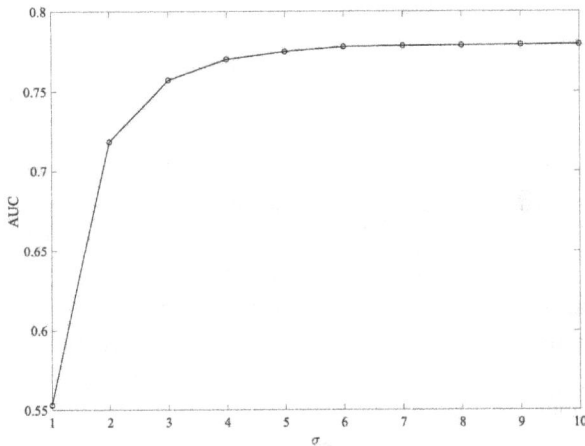

Fig. 2   The AUCs when λ, γ, k are set to $0.5, 0.025, 20$ and change σ from 1 to 10 with a step size of 1.

Since the area under the receiver operating characteristic curve (AUC) is independent of the threshold and the error, we use it as the performance metric. The value of AUC is $0 \sim 1$, and the greater the value, the better the performance of the model.

With the selected parameters we conduct a 5-folds cross-validation on other training and test sets. Seven reference methods which are LDA, LR, DT, NB, NN, SVM and RF are employed to compare the performance of the proposed method. The MATLAB functions corresponding to the seven reference methods are fitcdiscr(), fitglm(), fitctree(), fitcnb(), feedforwardnet(), fitcsvm() and TreeBagger() respectively, where feedforwardnet() is called with 10 hidden nodes, TreeBagger() with 90 trees, and other functions are directly run with the default options.

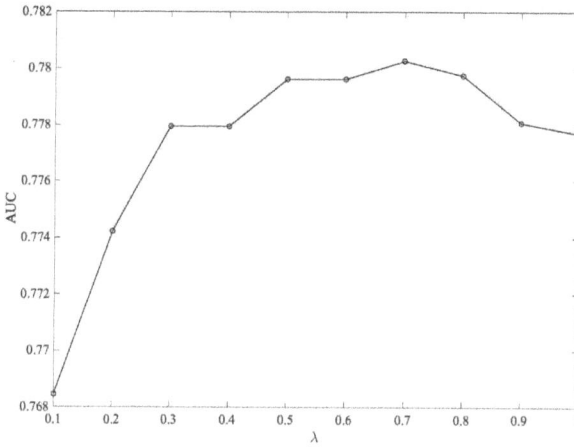

Fig. 3   The AUCs when σ, γ, k are set to $1, 0.025, 20$ and change λ from 0.1 to 1 with a step size of 0.1.

Table 1   Area under the receiver operating characteristic curve (AUC) results.

Method	Fold 1	Fold 2	Fold 3	Fold 4	Fold 5	Avg.	AR
Graph	0.7796	0.8089	0.8515	0.8394	0.8523	0.8263	1.0
LDA	0.7643	0.8014	0.8238	0.8177	0.8345	0.8084	3.0
LR	0.7649	0.8014	0.8255	0.8121	0.8303	0.8068	3.2
DT	0.6395	0.7502	0.7234	0.7025	0.7139	0.7059	7.8
NB	0.7412	0.7685	0.7989	0.7875	0.7641	0.7721	6.2
NN	0.7407	0.7027	0.7947	0.7971	0.7999	0.7670	6.8
SVM	0.7723	0.8085	0.8157	0.8101	0.8330	0.8079	3.2
RF	0.7439	0.7935	0.7987	0.8161	0.8296	0.7964	4.8

Friedman test statistic: $\chi^2_F = 30.87$, p-value$= 6.579e^{-05} < 0.005$

The results are list in Table 1, in which the AUCs of all the eight methods on each fold are listed. The method achieving the highest AUC on each fold is underlined. The experimental results show that the AUCs of the proposed method wins out over all the benchmark methods on each fold. The average AUC is as high as 0.8263, which is 17.9% higher than the second best method. The ROC curves of all the eight methods on the first fold have been plotted in Figure 6.

The average ranking (AR) on the five folds for each method is also listed, please refer to the last column of Table 1. AR is calculated as follows:

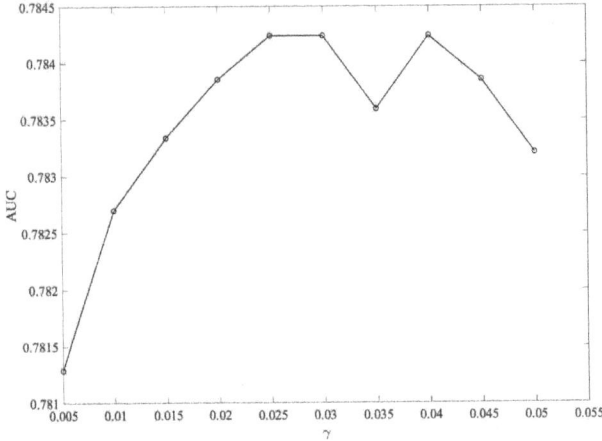

Fig. 4 The AUCs when σ, λ, k are set to $1, 0.5, 20$ and change γ from 0.005 to 0.05 with a step size of 0.005.

$$AR = \frac{1}{D} \sum_{i=1}^{D} r_i, \qquad (9.18)$$

where D denotes the number of folds and r_i is the rank of method on the ith fold. The AR of the proposed method equal to 1, which indicates that the performance of the proposed method ranks first in each fold.

We use Friedman's test [16] to compare the AUCs of the different models. The Friedman test statistic is based on the average ranking (AR) performances of the methods on each fold, and is calculated as follows:

$$\chi_F^2 = \frac{12D}{K(K+1)} \left[\sum_{j=1}^{K} AR_j^2 - \frac{K(K+1)^2}{4} \right], \qquad (9.19)$$

where K is the total number of methods and D denotes the number of folds. χ_F^2 is distributed according to the Chi-square distribution with $K - 1$ degrees of freedom. If the value of χ_F^2 is large enough, then the null hypothesis that there is no difference between the methods can be rejected. The last row of Table 1 shows the value of χ_F^2 is 30.87 and the $p-$value equal to $6.579e^{-05} < 0.005$, which indicates that ARs have statistical significance.

In order for the percentage reduction in the bad observations, we under-sampling the bad observations that only $233, 175, 124, 78, 37, 18$

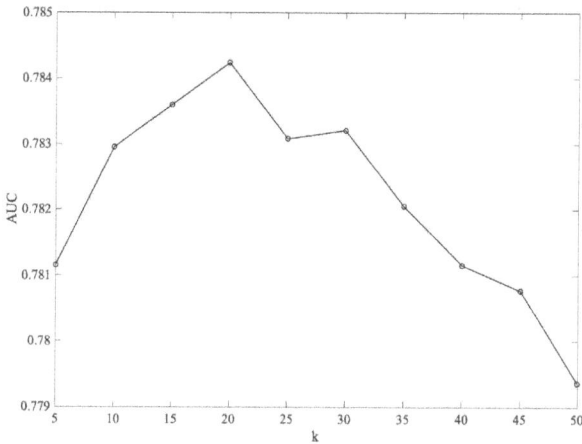

Fig. 5 The AUCs when σ, λ, γ are set to $1, 0.5, 0.025$ and change k from 5 to 50 with a step size of 5.

Table 2 The average of AUCs on five folds for the different percentage class splits.

Method	Percentage of class splits						
	70/30	75/25	80/20	85/15	90/10	95/5	97.5/2.5
Graph	0.8263	0.7894	0.7817	0.7601	0.7517	0.7758	0.6511
LDA	0.8084	0.7837	0.7865	0.7595	0.7272	0.7820	0.7407
LR	0.8068	0.7840	0.7816	0.7569	0.7307	0.7811	0.6588
DT	0.7059	0.6591	0.6328	0.6483	0.5449	0.5251	0.4993
NB	0.7721	0.7536	0.7600	0.7362	0.7047	0.7482	0.6908
NN	0.7670	0.7295	0.7443	0.6642	0.6677	0.5636	0.7573
SVM	0.8079	0.7788	0.7738	0.6971	0.6417	0.6616	0.5386
RF	0.7964	0.7690	0.7555	0.7353	0.6789	0.6991	0.5216

ones from a total of 300 bad observations have been used to give $75/25, 80/20, 85/15, 90/10, 95/5, 97.5/2.5$ class distributions. For each split, the 5-folds cross-validation is conducted, AR and the average of AUCs are calculated, the results are displayed in Table 2 and Table 3, in which the best results have been underlined. It shows that the proposed model is superior to the benchmark models on almost all the class splits. The averages of AUC are larger and the ARs are less than those of the benchmark methods.

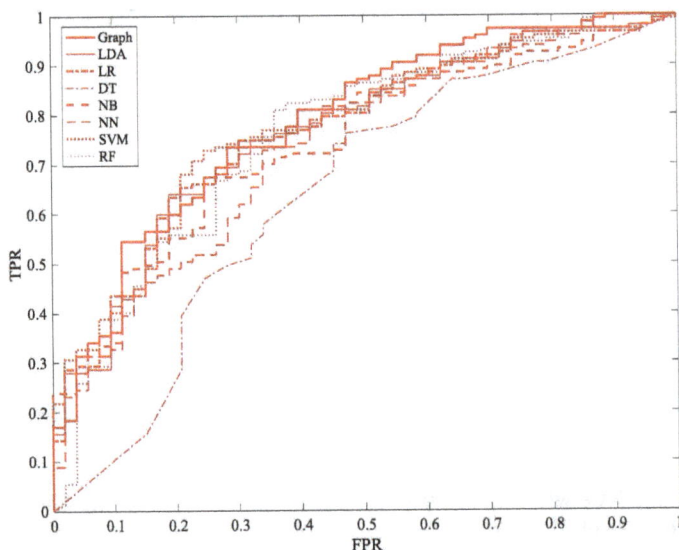

Fig. 6 The ROC curves of all the eight methods on the first fold. The abscissa represents the false positive rate (FPR), and the ordinate represents the true positive rate (TPR).

Table 3   ARs on the different percentage of class splits.

Method	Percentage of class splits						
	70/30	75/25	80/20	85/15	90/10	95/5	97.5/2.5
Graph	1.0	2.6	3.2	2.6	1.6	3.6	3.4
LDA	3.0	3.2	2.4	3.0	3.4	2.6	3.0
LR	3.2	3.2	3.4	2.8	2.2	2.2	4.4
DT	7.8	7.8	8.0	6.0	8.0	7.4	6.2
NB	6.2	5.6	4.4	4.0	4.6	3.8	4.0
NN	6.8	5.8	5.2	7.0	5.2	7.0	3.0
SVM	3.2	3.2	4.0	6.0	6.2	5.2	5.8
RF	4.8	4.6	5.4	4.6	4.8	4.2	6.2
χ^2_F	30.87	18.73	17.6	16.17	25.53	21.2	10.58
p-value	$6.579e^{-05}$	0.0091	0.0139	0.0212	0.0006	0.0035	0.1578

4   Conclusion

In this chapter, a new credit scoring method is proposed. To make full use of the information of unknown users to participate in the prediction, some novel strategies is figure out, they form the main innovations of the proposed method. We carry out experiments on the open UCI German credit dataset and compare them with seven classical models. Experimental results confirm these novel strategies have generated a very high prediction accuracy. The average AUC is as high as 0.8263, which is 17.9% higher than the second best method. The AR of the proposed method always ranks first in each fold and Friedman's test indicates that ARs have statistical significance.

Achnowledgements

This research was partially supported by NSFC (Nos. 11771458, 11601346) and Guangdong Province Key Laboratory of Computational Science at Sun Yat-sen University (2020B1212060032).

References

[1] W. E. Henley and D. J. Hand, Construction of a k-nearest neighbor credit-scoring system, *IMA Journal of Management Mathematics*, **8**, 4, pp. 305–32 (1997).

[2] B. Baesens, T. Van Gestel, S. Viaene, M. Stepanova, J. Suykens and J. Vanthienen, Benchmarking state-of-the-art classification algorithms for credit scoring, *Journal of the Operational Research Society*, **54**, 6, pp. 627–635 (2003).

[3] BrownI and C. Mues, An experimental comparison of classification algorithms for imbalanced credit scoring datasets, *ExpertSystems with Applications*, **39**, 3, pp. 3446–3453 (2012).

[4] G. Arminger, D. Enache and T. Bonne, Analyzing credit risk data: A comparison of logistic discrimination, classification tree analysis, and feed forward networks, *Computational Statistics*, **12**, pp. 293–310 (1997).

[5] S. Lessmann, B. Baesens and H. V. Seow, Benchmarking state-of-the-art classification algorithms for credit scoring: an update of research, *European Journal of Operational Research*, **247**, 1, pp. 1–32 (2015).

[6] D. West, Neural network credit scoring models, *Computers & OperationsResearch*, **27**, 11-12, pp. 1131–1152 (2002).

[7] B. Baesens and J. Vanthienen, Benchmarking state-of-the-art classification algorithms for credit scoring, *Journal of the Operational Research Society*, **54**, 6, pp. 627–635 (2003).

[8] L. Nani and A. Lumini, An experimental comparison of ensemble classifiers for bankruptcy prediction and credit scoring, *Expert Systems with Applications*, **36**, 2, pp. 3028–3033 (2009).

[9] O. Chapelle, B. Schölkopf and A. Zien, *Semi-supervised learning*. MIT Press (2013).

[10] M. Belkin, P. Niyogi and V. Sindhwani, Manifold regularization: A geometric framework for learning from examples, *J. Mach. Learn. Res.*, **7**, pp. 2399–2434 (2006).

[11] X. Liu, D. Zhai, D. Zhao, G. Zhai and W. Gao, Progressive image denoising through hybrid graph Laplacian regularization: A unified framework, *IEEE Trans. Image Process*, **23**, 4, pp. 1491–1503 (2014).

[12] D. Shuman, S. Narang, P. Frossard, A. Ortega and P. Vandergheynst, The emerging field of signal processing on graphs: Extending high-dimensional data analysis to networks and other irregular domains, *IEEE Signal Process. Mag.*, **30**, 3, pp. 83–98 (2013).

[13] A. Sandryhaila and J. M. F. Moura, Discrete Signal Processing on Graphs, *IEEE Transactions on Signal Processing*, **61**, 7, pp. 1644–1656 (2013).

[14] Xiaowen Dong, Dorina Thanou, Pascal Frossard and Pierre Vandergheynst, Laplacian matrix learning for smooth graph signal representation, *In: 2015 IEEE International Conference on Acoustics, Speech and Signal Processing (ICASSP)*, pp. 3736–3740 (2015).

[15] A. P. Bradley, The use of the area under the ROC curve in the evaluation of machine learning algorithms, *Pattern Recognition*, **30**, 7, pp. 1145-1159 (1997).

[16] M. Friedman, A comparison of alternative tests of significance for the problem of m rankings, *Annals of Mathematical Statistics*, **11**, 1, pp. 86–92 (1940).

[17] L. Xianming, Z. Deming, Z. Debin, Z. Guangtao and G. Wen, Progressive image denoising through hybrid graph laplacian regularization: a unified framework, *IEEE Transactions on Image Processing*, **23**, pp. 1491–1503 (2014).

[18] M. Belkin, P. Niyogi and V. Sindhwani, Manifold regularization: A geometric framework for learning from labeled and unlabeled examples, *Journal of Machine Learning Research*, **7**, pp. 2399–2434 (2006).

[19] Benjamin Girault, Shrikanth Narayanan, Antonio Ortega, Paulo Goncalves and Eric Fleury, GRASP: A Matlab Toolbox for Graph Signal Processing, *In: 2017 IEEE International Conference on Acoustics, Speech and Signal Processing (ICASSP)*, Mar 2017, New Orleans, United States (2017).

https://doi.org/10.1142/9789811239014_0010

Chapter 10

Optimal Choices of Features in Image Analysis

Camille Kurtz and Nicole Vincent

Université de Paris, LIPADE, F-75006, Paris, France
camille.kurtz@u-paris.fr, nicole.vincent@math-info.univ-paris5.fr

Abstract: This document discusses about the notion of features in image analysis. With feature we designate here some information issued from the raw image data, related to a visual pattern, that is supposed to have some interest for a particular task from the pattern recognition pipeline. After discussing about the *what for, whom for and how*, we present some examples from the literature highlighting the high diversity of features and some of their properties, which can be worth considering both by the computer scientist and the expert when designing an image analysis system. We finally provide some potential take-away messages to the reader, arguing that an optimal choice of features is rarely handleable since such a choice is more often a compromise between computation and application needs.

Keywords: Image analysis, Primitives, Hand-crafted, Features by learning, Feature choice, Multi-source

1 Introduction

During many years, in order to achieve any application in the fields of pattern recognition or image analysis, the researchers have worked a lot to build *features* that are efficient. According to the fields, the nature of the acquired data that has to be processed may differ. Here, we will concentrate our discussion on image analysis but a generalization is most often obvious.

If an image can be seen as a whole entity by the human, who is able to focus on a region of its interest, from a computer point of view it is an array of numerical values issued from a sensor figuring a spatially discretized reality. Depending on the sensor, different aspects can be measured such as color, depth, attenuation coefficient, that are later discretized, to define pixels. The pixel is then the atomic primitive but it often carries too local and elementary piece of information. In various applications, this is generally not sufficient to directly come to a conclusion.

In addition, images are often used as an intermediary medium that is built by experts of any domain. These experts try to construct images to get some intuition to perform their task and they are easier to work on than more abstract structures. This is the case for example in the field of time series [1], speech [2] or text analysis [3].

In the numerous papers from the image analysis domain, different terms are used such as characteristic, descriptor, representation or even, more recently, code. With feature, they are all designating some information issued from the raw data, feature with a more structural notion behind. A characteristic is usually issued from a calculus whereas a feature is more linked to a visual pattern. A descriptor is more used when a family of values computed from a single approach are concerned. Representation is also employed when indexing images to enable some future application such as content based image retrieval. The term code is considered when a set, ordered or not, of characteristics that have lost their semantic content are gathered in order to replace the raw data thanks to an embedding. In this paper, we consider the term *feature* as a generic term to designate some partial property associated with the considered item that can cover the other terms. It can be either numerical or symbolic.

Features are involved all along the parts of a global pattern recognition system. The later is usually composed of two blocks. In its design, first is the choice of the features and second is the decision-making part. More recently, the modern systems tend to be end-to-end, passing directly from the input data to an output decision. Then what happened to features? Should we change feature definition? We assume a feature is some computation from the raw data that is supposed to have some interest for a particular task, that is to say, that carries some interesting information or that enables to make easier the decision making. Features constitute a change in the nature of the data available to solve a considered problem.

We propose in this document to analyze the motivations to define features, to observe their variety, and how how they can be improved. Finally, we come to criteria enabling to make some choices according to the context.

2   A feature: what for, whom for and how?

In the Collins dictionary, it can be read *a feature of something is an interest-ing or important part or characteristic of it*. In the same way, in computer vision, a feature is either more relative to an abstract property that takes some values in a particular context or more physical when structural ele-ments are considered. The features are relative to the visual elements of the image. The main problems that happen in image analysis are linked to the segmentation of an image to get visual entities with semantic meaning, this can be to extract an object or a particular region of interest. But this can also be to affect some label or index to the image in a retrieval process, or to analyze the image from some point of view. Other domains are im-age restoration, image registration or problems that involve several images to detect changes in crisis event for example, or environmental changes in satellite image time series.

To build an answer to these problems, algorithms have to be proposed and systems to be implemented. In a very classical way, three different steps can be distinguished, the pre-processing, the processing and the post-processing ones. In all of them, some computation is done, based on in-formation. When solutions are described, the emphasize is put on the processing step where the features are well described. Nevertheless, they are not computed from the raw information but from the modified data through the pre-processing step. In the post-processing, features are also used, but most of the time their aim is to regularize the answer of the pro-cessing step. The features are no more computed on the initial image but on an approximation of the answer. Indeed some assumptions are done on the property of the exact solution and the aim is this property holds for the system answer.

In the whole system, several algorithms can be concatenated. An al-gorithm is a sequence or an organization of calculus that can associate an output to an input, it models a function. The feature in itself is also as-sociated with an algorithm. Its aim is not to get the final answer to the problem but it happens to be an intermediate process, a proxy, the result of which will be involved in a higher level process. In the process beginning, the feature is aimed at getting values, numerical or symbolic, that will gen-erally replace (or complete) the pixel values that are the elementary native features associated with the image. They can model human perception or simply enable to build a reasoning process between the image and the final result. Features can be used to progressively simplify the reasoning, by reducing the number of parameters of the problem. Another type of features are those aiming to catch correlations between expected results and input image.

Now, we can wonder how features are defined. Different points of view can be considered as a software is always motivated by the automation of a concrete task, image restoration, segmentation, analysis or classification for the most frequent ones. Previously we have analyzed the structure of the final software, but we can distinguish between different steps in the building of a solution that is not the work of a lonely expert. It begins with the expression of the need of an applicant. Then the design of an algorithm as a theoretical solution has to be proposed and finally, the effective software has to be implemented taking into account the constraints expressed by the applicant, these may be associated with computation time or hardware requirements (e.g. embedded systems, or GPU / TPU for deep learning).

Then, two approaches can be differentiated. We analyze here these different visions and their associated constraints:

- In the computational approach, the structure of the data on which the reasoning will be based is an important question. This structure is then linked to the mathematical tools that are available to manipulate the data according to its properties (e.g. distribution). Most of the problems are solved through an optimization process. The main theorems and tools can be applied in vector spaces, for strings or graphs. Then, the data are to be modeled as vectors, strings or graphs;

- The expert has his own point of view, he knows the result that we can expect from the initial image; at least, he has one answer that he considers the only solution. Even if the expert has some difficulty to express its reasoning mode, he looks at the image and has a conclusion according to some elements (features structural or not) he is evaluating (numerically or symbolically). The right property is considered. In some cases, the expert is not aware of the exact reason of its conclusion. One of the expert aim, can be to express some correlation that he is not aware of (e.g. a biomarker), or to express the features that are associated with a clustering he is satisfied with.

From the previous points of view and as very often opposed, some features are said *hand crafted* when they are computed from a non iterative or aleatory computation and *automatic* when the feature value is depending

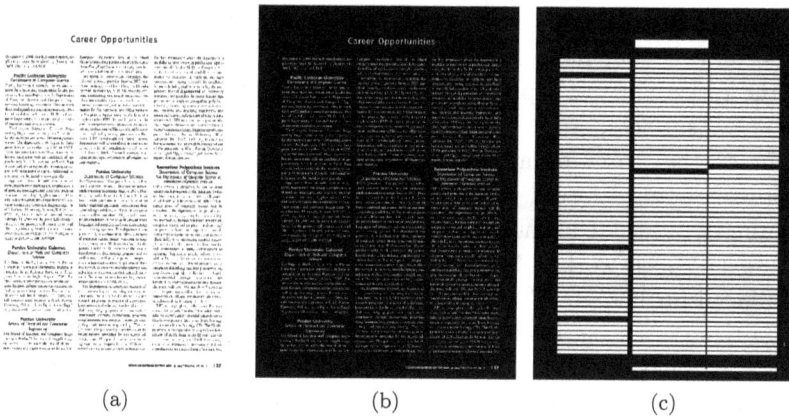

Fig. 1   Extraction of straight line segments from a binary document (a) to approximate characters (b), or lines (c) in a document layout analysis process [6].

on contextual data (e.g. learning data). Nevertheless this distinction is not always well-justified, as we can say all features are handcrafted [4]. Indeed the architecture of the algorithm, that has to be used iteratively and on numerous data, has to be manually written and implemented by the computer scientist. In the case of deep learning process, the best architecture has to be invented and actually no concrete way to learn the best architecture is available.

3   Diversity of features

We have discussed the context of defining features, we now are to discuss the types of features most often used, highlighting their diversity.

3.1   *Data structure choice: a mathematical point of view*

Known mathematical theorem being the core of any reasoning process, the first choice concerns the type of space in which the image representation is considered. The most frequent is a vectorial space of any dimension. Sets (or bags) with less structuring properties are possible but contain some structured data. Strings and graphs (including trees) are also representations used in different contexts, for instance for the indexation of temporal sequences [5].

Choice of primitives   In many approaches, the primitive corresponds to the pixel ($0D$) but more sophisticated ones can be considered, depending on the tasks and the applications, for example $1D$ straight lines, circles, $2D$ or $3D$ surfaces, etc.

In [6], the chosen primitives are $1D$ straight line segments, enabling to approximate filled forms, lines and drawings that constitute another level of primitives in document analysis from a topological point of view (see Figure 1). Such work is based on the Local Diameter Transform (LDT) mentioned further.

In remote sensing image analysis, the preferred primitives often correspond to homogeneous $2D$ regions of the images, providing more compact and high-level sets of primitives to deal with [7].

Representation structures   The structure chosen has nothing to do with the features considered. A vector space may contain as components some various independent features or just be the result of a change of basis in the matrix representation of the image. This is the case of the Fourier transform, DCT or all the family of wavelets [8].

Another example is a graph that can be graph of relative position of the elements extracted in the image for a structural description of the image [9]. But it can also be as a hierarchical structure of the features in order to highlight some properties enabling to discriminate between the patterns to recognize. In [10], a framework to build binary partition trees (BPT) is proposed. A BPT allows for a hierarchical representation of an image in a multiscale way, by providing a tree of nodes corresponding to image regions (see Figure 2). In particular, cuts of a BPT can be interpreted as segmentations of the associated image. Building the BPT of an image then constitutes a relevant preliminary step for optimization-based segmentation methods and / or object detection tasks.

First-order features   Color, shape and texture are the most common motivations to define a feature.

The most simple is the **color** of each pixel in the image. The color image histogram is a descriptor that provides not so bad results in content-based image retrieval [11]. The histogram is a $1D$ vector and its dimension can be chosen according to the need. Besides, it can be interpreted when needed as a distribution.

As **shape** descriptors we mention the descriptors associated with the transforms linked to a change of basis. The descriptor is made of a set

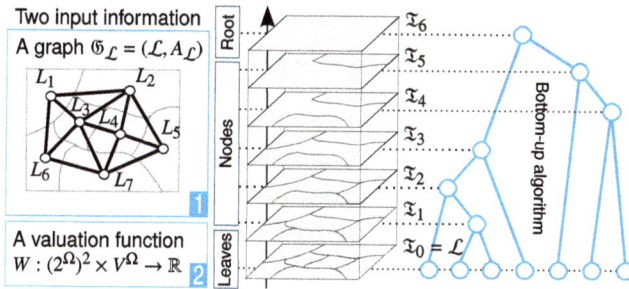

Fig. 2 A binary partition tree (BPT) provides a graph-based hierarchical representation of an image in a multiscale way [10].

of coefficients generally structured in a $1D$ vector, the size of which has to be chosen as a parameter of the method. Many studies have multiply these descriptors so that they have specific properties. Fourier transform provides Fourier Descriptor (FD). It has been improved as Generic Fourier Descriptor (GFD) [12] with better invariance properties. Wavelet descriptor (WD) are defined as well as moment descriptors such as Geometric Moment Descriptor often used as Hu Moments, or Zernicke Moment Descriptors (ZMD) used for example for character recognition.

These moments can also be used as **texture** descriptors, used for texture segmentation purpose. More recently, the Local Binary Patterns (LBP), gathered in histograms, are well-known texture features that have been used with success in the face recognition domain [13].

In [6] a new transform, called the Local Diameter Transform (LDT), is proposed to integrate the local spatial organization of the segments, chosen as primitives. The image pixels are then characterized by both the orientation and the lengths of the maximal directional segment crossing the pixel, such features being discriminative for segmentation tasks.

Feature can be computed at a **global** level of the image but it can also be computed **locally**. The color is a local feature, made global thanks to the histogram.

The most local one is surely directional gradient computed from the Roberts Kernel. It can be computed on every pixel of the image thanks to a convolution operation between the image and the kernel. The possible kernels are numerous and have different properties to smooth images. Such features, computed locally, are generally used as a modification of the initial image.

Some more sophisticated local features have been proposed, mainly in the context of object tracking or indexing. The intensity and the direction of the color variations around a point is quite characteristic of a zone, both shape and texture, often independent of the intensity and contrast. This led to Histogram of Oriented Gradient (HOG) computed has concatenation of histograms of the direction of gradients, in a large neighborhood of a pixel. Scale-Invariant Feature Transform (SIFT) is usually used in junction with key-point extraction to characterize the key-point. Its definition has been motivated by the need to get characteristics invariant to scale. Then the neighborhood of the point is studied at different scales and the best scale is selected. In this trend, many transforms were proposed such as the Oriented fast and Robust Brief (ORB) [14], and Learned Invariant Feature Transform (LIFT) [15] to cite only two.

The features issued from a deep Convolutional Neural Network (CNN) result from a learning process, they are optimized in an end-to-end process to solve the task. Such features derive from linear convolutions on raw color pixels, combined with transfer functions and pooling steps, orchestrated in a multilayer process. Different levels of features can be obtained according to the layer (from low levels, capturing local patterns such edges, to more high levels related to object sub-parts), and we can consider them in a hierarchical dependency. Then, they are related to the application (segmentation or classification task) and to the learning dataset on which they have been optimized. The transfer learning strategy enables to use some optimized features in other context where they happen to give better results than new features and they can also be modified to be better adapted to some specific data. In [16], transfer learning is used to overcome the few samples generally available in medical contexts and applied to different handcrafted image representations of patients drawings in a diagnosis task.

Beside these classical features, Mandelbrot expressed in 1977 that Euclidean approaches were only an approximation and that the nature was fractal [17]. The observation at one scale gives only a partial vision and the important thing is to measure the modification when observation scale is varying. A modeling thanks to power laws is then more suited [18]. This is in accordance with the visual human perception that is prone to like monuments where the same patterns are present at different scales (e.g. cathedrals portal). There is an important difference with the deep neural networks where the reception fields of the different layers are getting larger and larger from the input to the output. Indeed, only the last vision is used in the decision step. Nevertheless some systems concatenate the output of several layers as input of the decision step. The fractal approach enables

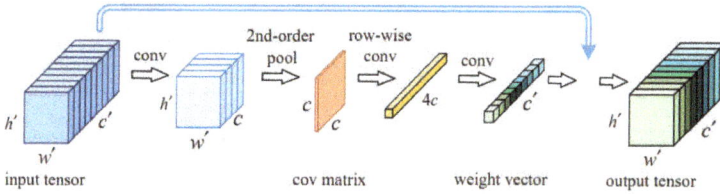

Fig. 3 The GSoP block proposed in [24] to introduce sencond-oder information in a CNN architecture.

a global approach rather than the analysis of objects, parts difficult to be extracted as in composite materials [19, 20].

Second-order features First-order features presented previously have shown their interest in many image processing and computer vision applications. They can have good properties and have the main advantage of being quick to calculate. However due to their simplicity, they suffer from different limits.

For example in the context of texture analysis, such features provide most of the time information related to the gray-level distribution of the image, however they do not carry information about the (relative) positions of the gray levels through the image content. They are then unable to quantify whether all high gray level values are positioned together, or they are twisted with the low gray level values.

To deal with this issue, second-order features have been often used in computer vision, via for example the notion of Region Covariance Descriptors (RCDs) [21], encoding covariance matrices computed from local image features. The co-occurrence matrix is defined over an image and models the distribution of co-occurring pixel values in a given window. Such structure has been used largely for texture analysis, with various applications in medical image analysis [22]. Since co-occurrence matrices are often large / sparse, various features are computed over the matrices, such as the Haralick features. Second-order statistics computed from first-order (handcrafted) features have proven relatively powerful in various image recognition tasks, in line with Psychophysics research that shown they play a particular role in the human visual recognition system [23].

In the context of deep learning and CNNs, some recent architectures were proposed to learn more discriminative representations, closer to second-order features. Indeed, classical networks, by performing linear combinations and elementary nonlinear operations, can be considered as

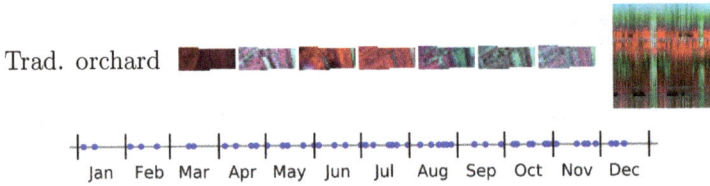

Fig. 4   Temporal evolution of a traditional orchard according (left) to its initial $2D + t$ representation from a SITS and (right) to a $2D$ spatio-temporal representation [31]. The last line shows the distribution of the images from the SITS (2017).

extracting only first order information from an input image. In [24], a method is proposed for capturing holistic statistical correlations through a CNN, by adding higher-order pooling steps into intermediate layers of the network (see Figure 3). Authors of [25] proposed the notion of Covariance Descriptor Unit (CDU) to replace the fully-connected layers of standard CNNs, integrating second-order information for image recognition.

A higher level of image description may be defined by studying the spatial organizations of pairs of objects present in an image content (with the model of human perception [26]). Taking into account spatial relations may then improve various pattern recognition applications such as scene recognition, or content-based retrieval approaches, when dealing with highly spatially structured images. Spatial relation features can complement efficiently the traditional features based on outer contour, geometry, texture, color, or even deep convolutional features, that are not always adapted to successfully describe image contents composed of objects mutually arranging with complex spatial configurations. In this context we can cite the force histograms [27] that capture well generic spatial relations. By computing the interaction between objects for each direction, it combines the directional aspect and distance measures, while taking into account the objects dimensions. Some other descriptors were also proposed for more particular spatial relations, for instance for the relations *across*, *along*, *between* [28, 29], or *enlaced by* [30], with a strategy inspired from the force histogram for the latest. In this way, the relative position of an object with regards to another one can have a representation of its own, from which it is possible to derive rich features.

3.2   *Multisource*

In image analysis, it is relatively common to have several data sources providing information related to the same sensed scene. For example, in

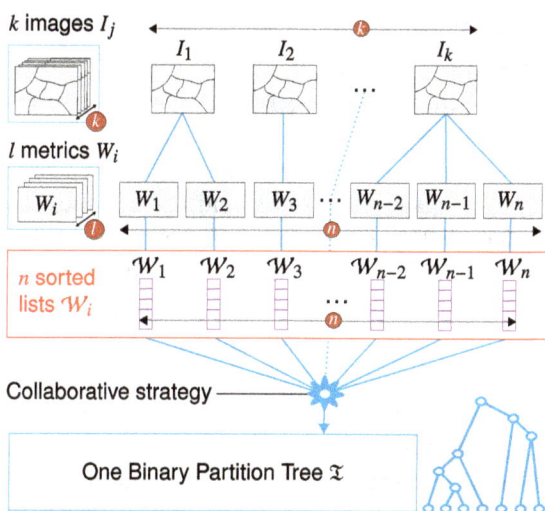

Fig. 5   BPT construction based on a consensus of features, potentially computed from multiple sources of images [10].

medical imaging, devices can simultaneously acquire CT and nuclear images of a same patient in a single exam. Note that, basically, a color image can also be seen as a multisource representation of a scene, since the different RGB channels do not carry the same information. From a smart-phone, you can capture an image of a document to scan it in burst mode, producing multiple acquisitions from slightly different angles. In remote sensing application, temporal series of images can represent the same spatial region. Furthermore, it is also possible to have textual information, keywords, accompanying the image. We often talk about multimodal data.

An other context leads to multisource when different families of features are considered, then some aggregation process has to be achieved. The notion can be observed when, first, sets of transforms leading to different images are available on which features are extracted. In any case, the question that arises is the interest of considering these different sources of information simultaneously in the recognition process. To allow the latter to benefit from it, the data must carry complementary information, not too redundant.

To deal with the different sources of information in image analysis, a fusion is required. It can be applied at different levels: (early) data level, (mid) feature level or (late) decision level. The choice of the fusion level generally depends on the application and the diversity of the different information sources. When the scene is represented by two raster images of similar natures, a early fusion at pixel level or a concatenation of the values of each source (image stacking) is possible. Classical image analysis algorithms relying on features can then be employed on this fused image. However, when the different images are not of directly comparable nature or when dealing with image data source and textual data source, a mid or late fusion is preferred. For mid-fusion, different features can be computed from the different sources and can be aggregated, leading to a common feature space in which reasoning becomes possible. The most important problem can come from the different units or scales associated with the feature values. They can be aggregated thanks to the concatenation of the single feature and it is the role of a similarity measure to tackle the differences. They can also be added when dealing with dual path U-Net architecture CNN. The use of a bag of features is another solution based on histograms, more sophisticated are the Vector of Locally Aggregated Descriptors (VLAD) [32] and Fisher Vectors approaches that can be applied to vector features. They are often employed when images are characterized by SIFT local features. For late fusion, features are classically computed from each data source leading to different feature spaces and a fusion is operated at the decision level, for example by voting on the decisions of different classifiers.

When dealing with multitemporal data such as satellite image time series, one concern is to mix spatial and temporal information to provide more rich information to the decision making process. A pre-processing process can be employed to reduce the dimension of the data. This is done in [31] using Hilbert-like space filling curves to build an intermediate image summarizing both temporal and spatial information from the original SITS (see Figure 4). Spatio-temporal features can then be learned from this representation by feeding a classical 2D CNN.

In [10], an approach is proposed to build a hierarchical image representation from multiple sources of information, as BPTs. Basically, a BPT is built for one given image, and one given feature, by merging pairs of connected image regions that are similar in the induced feature space. The proposed approach enables to manage multiple features within the BPT building process. It relies on a collaborative strategy used to establish a consensus between different metrics assessing feature similarities, thus enabling to obtain a unified hierarchical segmentation space (see Figure 5). In particular, this provides alternatives to the complex issue of metric construction from several –possibly non-comparable– features computed from

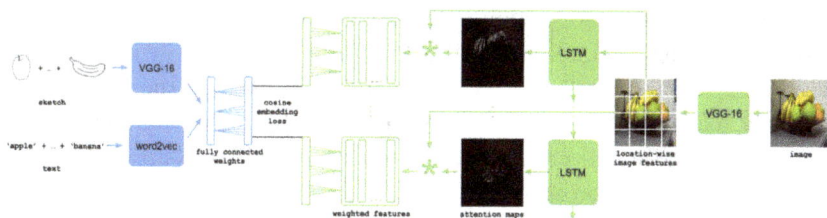

Fig. 6 Aggregation of features computed on image and on text [33].

(potentially multiple) image content(s).

In deep approaches, for example based on CNNs, we often find the same fusion patterns. If we have 3 grayscale images, we can stack them, to form a 3-channel image compatible with standard pre-trained CNNs (early fusion). We can also train networks on each data source then merge the codes associated with each CNN, then train a Multi-Layer Perceptron (MLP) or any other classifier (mid fusion). It is also possible to tweak an embedding, learned from one source, via another source. A late fusion in CNN decisions is also trivial.

A more hybrid level of fusion is also possible. If two data sources are available, two independent CNNs can be connected together via a common layer or a common Loss function. Via back-propagation, the features learned in each of the two networks are then jointly dependent on both sources of information. If a common layer is present, it figures a common embedding space where distances are more stable. Then several family of features are considered and they are automatically fixed and made dependent according to the application (see Figure 6 from [33]).

4   Feature properties

Of course the only final property that is asked to features is to enable to achieve a task with the highest possible accuracy. Nevertheless we can distinguish different families of properties that have been highlighted along time and with respect to different applications.

Invariance properties   Feature invariance has been an active research area of image analysis. In itself, the term has no meaning, the invariance has to be with respect to something. By invariance is meant not only features that are invariant to a set of geometric transforms, but also methods for carrying out pattern recognition independently not only of the position and pose but to imaging conditions using, features that are not invariant themselves. The invariance requirements generally depends on the application. In object recognition task, for example, orientation, scale, illumination and translation invariances are the main requirements. Databases have been built to check this invariance.

When the computer scientist involves a *hand-crafted* feature in a pattern recognition process, he/she must be careful to take these properties into account. In general, such properties are well-known for classical features and have been demonstrated in the literature. For example, the Zernike moments are natively rotation-invariant features, while scale and translation invariances can be obtained by normalizing the image using its regular geometrical moments [34]. By definition, the LBPs are only invariant to monotonic light changes in gray-level, but some researches have been done to make them invariant in rotation as well as to increase their photometric invariance property, see [35] for a survey.

When a novel feature is designed, the computer scientist must prove invariance properties theoretically, or at least test them experimentally, to allow their re-use in other applications by the community. In the context of deep features, such as convolutional ones, the study of invariance properties is much more complex since they are directly learned by the neuronal networks and the properties result from the considered architectures [36], as well as from the learning set. Convolutional features have shown to model well the local appearance of natural images at multiple scales, while also representing images with some invariance through the numerous pooling operations. However, the theory of this invariance and the properties of the embedding spaces built by the CNNs are weakly manageable. Most of the time, these models are seen as *smart* black boxes, accompanied with many insights recorded since they were introduced. Nevertheless, [37] claims some strategies to learn invariant feature hierarchies.

Finally, every pattern recognition system based on learning aims to extend a function known on a training set. The invariance to the features with respect to the label cannot be considered outside of the global system. The invariance to the label is then called the **generalization** ability of the system.

Semantic properties   What is semantics? Indeed, this a very subjective notion. How does an image have semantics? It has many semantic contents, depending on the observer. Generally, we use the term to designate some visual thing that can be described by words. But it can be considered at different levels. A pixel can be dark or a region can be round or can be a face. The lowest level is pixel level and the highest level is linked to expert's expectations, both being separated by the *semantic gap*, and for a single process, different semantic levels can be considered.

In some applications, domain knowledge may be employed to bridge this semantic gap, for example in the domain of content-based image retrieval. The domain knowledge must however be structured to make it useful in a computer system. In [38], images are characterized by vectors of semantic terms, and a similarity measure that integrates semantic and visual relations between terms, is employed to retrieve similar images. In deep approaches the semantic aspect is generally looked for after the problem is solved, by visualization of the different filter layers, the interpretation is not always clear.

Interpretability properties In many cases, particularly in industrial or medical applications, the end-user asks to know why the decision is taken. Reasoning mode and features are involved in the global interpretability. The computer scientist point of view is different. To get an interpretable feature he will introduce an algorithm that might not be understood by the end user. This makes the difference between semantic and interpretability property. For example *linearity* of a writing is a semantic feature but can be measured by different algorithms that are not all interpretable in the process.

Locality properties Reference to the domain of influence of a feature is often made. Some features involve only pixels in a limited region. This is the case of SIFT that are both local and invariant to scale. On the opposite Fourier transform leads to Fourier coefficients involving all the image pixels. Most often, global features are difficult to interpret by end-users. Of course global features can be computed more locally. We can classify Hough transform in semi local approaches.

Expressivity properties They are related to the efficiency of the features with respect to a goal, discrimination of classes or capacity to specify a class. This is particularly important in biomedical applications were bio-markers are looked for. Here they are supposed to have a causality effect with a pathology, at least a high correlation. Features may appear as computer-markers of an object or a class.

Operational constraints   When considering a descriptor or a set of features and in some applications, some operational constraints may appear such as time constraints, storage capacity or security constraints. Then, the complexity of the algorithm to compute feature has to be taken into account. Some compromise between time and accuracy has to be established according to the context. For example, in mobile applications or services, memory constraints lead to adapt the method by also considering energy saving objectives. Often, features are descriptive and the original data could be recovered in totality or partially, for example in medical or biometric imaging. In that case, confidentiality cannot be assured. This is another criterion to choose a set of features, compactness and sparseness for robustness.

The choices also strongly depend on the practice and culture of the designer of the systems. Besides, all these properties enable to build different taxonomies of features.

5   Selection versus Optimization

Whatever the approach to define features is used, the number of features is most often very large as the designer wants to get as much information as it can. Nevertheless, either some are redundant, others are non useful and others bring some noise when solving the problem in the new representation space. Besides in spaces with high dimension most of the mathematical results proposed are not efficient. This is known as the curse of dimensionality. Two main types of methods have been proposed to improve the efficiency of a set of features. Those that select a small set of features, keeping the information computed by the initial ones, and those that modify the initial features thanks to some combinations and decrease the number of finaly retained features.

Given a set of features that seem relevant for a particular image analysis task (according to the computer scientist or the expert point of view), it is generally necessary to operate a **selection** to choose those that will actually be used in the different phases of the pattern recognition process. For example, when analyzing a biological image of cells, the objects of interest can be segmented according to a colorimetric criterion and then classified according to geometric criteria.

Obviously, such a task can be carried out manually by the computer scientist, thanks to his image culture, or even by a trial-and-error process, or by the expert, thanks to his knowledge of the domain, but this process is tedious and may not lead to an optimal result.

Feature selection is a field of research in itself [39]. It aims to select a subset of relevant features to construct a model. The notion of relevance is related here to the discriminative power of the features for a particular problem and their non-redundancy. Such a selection strategy can be useful to simplify the models to make them more interpretable by the expert (and to deal with the curse of dimensionality for the computer scientist when a vector space is considered), to decrease the training times, to enhance generalization by reducing overfitting, etc. Feature selection techniques are often used in image analysis where numerous features are involved, while only few samples are available.

The main hypothesis is that the computed features from the images can be either redundant or irrelevant, and can thus be removed without losing too much information. One must be careful with the notions of redundancy and irrelevance, since a relevant feature can be redundant in the presence of another correlated relevant feature.

An algorithm for feature selection relies on a search technique for proposing new subsets of features, integrating a fitness measure (supervised or not) aiming to assess quantitatively the different subsets. An exhaustive search of the space is generally a non-tractable strategy. The fitness measure is the key element since it influences deeply the algorithm. Three main families of feature selection algorithms co-exist:

- the Filter methods rely on training data properties, independently of the task to be performed, they can use mutual information or correlation coefficient measures instead of the accuracy of the model to assess a feature subset. These methods do not take into account interactions between features and tend to select features that are redundant rather than complementary;
- the Wrapper methods evaluate feature subsets on the basis of their performances using a learning algorithm;
- the Embedded methods which are a batch of techniques embedding the feature selection directly inside the model construction process.

An illustrative example of the last family of algorithms can be found in [40] where a fast feature selection method is proposed. In this approach, each feature is closely associated with a single feature classifier. The weak classifiers considered have several degrees of freedom and are optimized on the training dataset. Within the genetic algorithm, the individuals who are classifier subsets are evaluated by a fitness function based on a combination of single feature classifiers.

Feature selection strategies should not be confused with feature **optimization** ones. While feature selection provides a feature subset, feature optimization (also referred as feature extraction in the machine-learning literature) aims to build new features as one or more functions of the original features. When designing an image analysis system, the objectives can be to build a novel feature space with not too many features, composed of uncorrelated features. Such novel features can be built by combining original features in an optimized manner. The combination approaches can be either linear or non-linear.

Principal Component Analysis (PCA) is a useful (unsupervised) linear transformation for doing that [41]. By determining the eigenvectors of a covariance matrix with the highest eigenvalues, PCA uses this information to project the features into a new subspace of lower dimension, capturing a significant part of the information found in the original features and defining new non-correlated features. Nevertheless the method only takes into account linear dependencies between features. Multi Dimensional Scaling (MDS) allows to build a low dimension representation space where the distances between samples are best preserved.

Non-linear methods aim to optimize the representation of the initial topology of the data. Isomap or Locally Linear Embedding (LLE) [42] are the most popular with the same objective than MDS.

Feature optimization can also be realized using supervised strategies. The Linear Discriminant Analysis (LDA) enables to find a linear combination of features that characterizes or separates two or more classes in a pattern recognition task [43]. As PCA, LDA works when the features values are continuous quantities.

These methods have been developed and are adapted to sets of hand-crafted features. Optimization is also extremely present in the deep approaches. Linked to the architecture, a large number of features belonging to a family of functions are computed and the parameters are optimized according to a loss function associated to the task. Dropout operations in the learning process enable to increase the generalization property of the system by limiting *by heart* learning. The optimization is performed for the definition and not to minimize the number of features. Nevertheless, a convolutional auto-encoder architecture tends to perform an ACP. Besides, the attention processes are added either to define new features or to limit the features to those that were the most adapted to the task [44].

Most of the methods we introduced are based on a learning phase, on a training set on which the optimization is performed. In industrial systems, the task can be relevant for a long time but the samples may evolve or in a classification task the number of classes may vary. Then, the features have to be adapted on line. Many methods have been proposed, often depending on the reasoning step nature or on the non stationary nature of data [45,46].

6 Conclusion

Any choice of features is difficult as it involves some criteria that can differ from an application to the other and from a designer to an other. Indeed, the solution is not unique and, when features are concerned, the choice is also depending on the choices done in the other parts of the entire pattern recognition process. Actually, the end-to-end learning processes let us think the problem does no longer exist as the features are automatically defined. In fact, the strategy is hidden in the architecture of the system and in the loss function to be optimized. In any case, the real word has to be modeled either with statistical or structural models. The best way to define features is to optimize some functions that can be intended to define a model or a global system. Mathematical tools for optimization have to be considered taking into account the non imperfection of the models that are often used. Many improvements are brought for example by introduction of regularization terms.

References

[1] Z. Wang and T. Oates, Imaging time-series to improve classification and imputation, in *IJCAI, Procs.*, pp. 3939–3945 (2015).

[2] O. Abdel-Hamid, A. Mohamed, H. Jiang, L. Deng, G. Penn and D. Yu, Convolutional neural networks for speech recognition, *IEEE/ACM Transactions on Audio Speech and Language Processing* **22**, 10, pp. 1533–1545 (2014).

[3] Y. Kim, Convolutional neural networks for sentence classification, in *EMNLP, Procs.*, pp. 1746–1751 (2014).

[4] N. Vincent and J. M. Ogier, Shall deep learning be the mandatory future of document analysis problems? *Pattern Recognition* **86**, pp. 281–289 (2019).

[5] J. Lin, E. J. Keogh, S. Lonardi and B. Y. Chiu, A symbolic representation of time series, with implications for streaming algorithms, in *DMKD workshop, Procs.*, pp. 2–11 (2003).

[6] H. Alhéritière, F. Cloppet, C. Kurtz, J. M. Ogier and N. Vincent, A document straight line based segmentation for complex layout extraction, in *ICDAR, Procs.*, pp. 1126–1131 (2017).

[7] T. Blaschke, Object based image analysis for remote sensing, *ISPRS Journal of Photogrammetry and Remote Sensing* **65**, 1, pp. 2–16 (2010).

[8] J. Z. Wang, Wavelets and imaging informatics: A review of the literature, *Journal of Biomedical Informatics* **34**, 2, pp. 129–141 (2001).

[9] A. Dutta, J. Lladós, H. Bunke and U. Pal, Product graph-based higher order contextual similarities for inexact subgraph matching, *Pattern Recognition* **76**, pp. 596–611 (2018).

[10] J. F. Randrianasoa, C. Kurtz, E. Desjardin and N. Passat, Binary partition tree construction from multiple features for image segmentation, *Pattern Recognition* **84**, pp. 237–250 (2018).

[11] G.-H. Liu and J.-Y. Yang, Content-based image retrieval using color difference histogram, *Pattern Recognition* **46**, 1, pp. 188–198 (2013).

[12] D. Zhang and G. Lu, Shape-based image retrieval using generic fourier descriptor, *Signal Processing: Image Communication* **17**, 10, pp. 825–848 (2002).

[13] Z. Chai, H. M. Vazquez, R. He, Z. Sun and T. Tan, Semantic pixel sets based local binary patterns for face recognition, in *ACCV, Procs.*, pp. 639–651 (2012).

[14] E. Rublee, V. Rabaud, K. Konolige and G. R. Bradski, ORB: an efficient alternative to SIFT or SURF, in *ICCV, Procs.*, pp. 2564–2571 (2011).

[15] K. M. Yi, E. Trulls, V. Lepetit and P. Fua, LIFT: learned invariant feature transform, in *ECCV, Procs.*, pp. 467–483 (2016).

[16] M. Moetesum, I. Siddiqi, N. Vincent and F. Cloppet, Assessing visual attributes of handwriting for prediction of neurological disorders - A case study on parkinson's disease, *Pattern Recognition Letters* **121**, pp. 19–27 (2019).

[17] B. Mandelbrot, *Fractal Geometry of Nature*. W. H. Freeman (1977).

[18] Y. Caron, P. Makris and N. Vincent, Use of power law models in detecting region of interest, *Pattern Recognition* **40**, 9, pp. 2521–2529 (2007).

[19] H. Boulecane, N. Vincent, M. Ruffier and H. Emptoz, Control of composite material structure by fractal methods, in *CAIP, Procs.*, pp. 726–731 (1993).

[20] A. Akrami, N. Nasiri and V. Kulish, Fractal dimension analysis of mg2si particles of al-15mg2si composite and its relationships to mechanical properties, *Results in Materials* **23**, pp. 100–118 (2020).

[21] O. Tuzel, F. Porikli and P. Meer, Region covariance: A fast descriptor for detection and classification, in *ECCV, Procs.*, pp. 589–600 (2006).

[22] L. Nanni, S. Brahnam, S. Ghidoni, E. Menegatti and T. Barrier, Different approaches for extracting information from the co-occurrence matrix, *Plos One* **8**, 12, pp. 1–9 (2013).

[23] B. Julesz, E. N. Gilbert, L. A. Shepp and H. L. Frisch, Inability of humans to discriminate between visual textures that agree in second-order statistics–revisited, *Perception* **2**, 4, pp. 391–405 (1973).

[24] Z. Gao, J. Xie, Q. Wang and P. Li, Global second-order pooling convolutional networks, in *CVPR, Procs.*, pp. 3024–3033 (2019).

[25] K. Yu and M. Salzmann, Second-order convolutional neural networks, *CoRR* **abs/1703.06817** (2017).

[26] G. D. Logan and D. D. Sadler, A computational analysis of the apprehension

of spatial relations, in *Language, speech, and communication. Language and space*, pp. 493–529 (1996).

[27] P. Matsakis and L. Wendling, A New Way to Represent the Relative Position between Areal Objects, *IEEE Transactions on Pattern Analysis and Machine Intelligence* **21**, 7, pp. 634–643 (1999).

[28] I. Bloch, O. Colliot and R. M. Cesar, On the Ternary Spatial Relation "Between", *IEEE Transactions on Systems, Man, and Cybernetics, Part B: Cybernetics* **36**, 2, pp. 312–327 (2006).

[29] N. Loménie and D. Racoceanu, Point set morphological filtering and semantic spatial configuration modeling: Application to microscopic image and bio-structure analysis, *Pattern Recognition* **45**, 8, pp. 2894–2911 (2012).

[30] M. Clément, A. Poulenard, C. Kurtz and L. Wendling, Directional Enlacement Histograms for the Description of Complex Spatial Configurations between Objects, *IEEE Transactions on Pattern Analysis and Machine Intelligence* **39**, 12, pp. 2366–2380 (2017).

[31] M. Chelali, C. Kurtz, A. Puissant and N. Vincent, Image time series classification based on a planar spatio-temporal data representation, in *VISAPP, Procs.*, pp. 276–283 (2020).

[32] H. Jégou, M. Douze, C. Schmid and P. Pérez, Aggregating local descriptors into a compact image representation, in *CVPR, Procs.*, pp. 3304–3311 (2010).

[33] S. Dey, A. Dutta, S. K. Ghosh, E. Valveny, J. Lladós and U. Pal, Learning cross-modal deep embeddings for multi-object image retrieval using text and sketch, in *ICPR, Procs.*, pp. 916–921 (2018).

[34] A. Khotanzad and Y. H. Hong, Invariant image recognition by zernike moments, *IEEE Transactions on Pattern Analysis and Machine Intelligence* **12**, 5, pp. 489–497 (1990).

[35] L. Nanni, A. Lumini and S. Brahnam, Survey on LBP based texture descriptors for image classification, *Expert Systems with Applications* **39**, 3, pp. 3634–3641 (2012).

[36] W. Liu, Z. Wang, X. Liu, N. Zeng, Y. Liu and F. E. Alsaadi, A survey of deep neural network architectures and their applications, *Neurocomputing* **234**, pp. 11 – 26 (2017).

[37] Y. LeCun, Learning invariant feature hierarchies, in *ECCV, Procs.*, pp. 496–505 (2012).

[38] C. Kurtz, A. Depeursinge, S. Napel, C. Beaulieu and D. L. Rubin, On combining image-based and ontological semantic dissimilarities for medical image retrieval applications, *Medical Image Analysis* **18**, 7, pp. 1082–1100 (2014).

[39] I. Guyon and A. Elisseeff, An introduction to variable and feature selection, *Journal of Machine Learning Research* **3**, pp. 1157–1182 (2003).

[40] H. Chouaib, F. Cloppet and N. Vincent, Combination of single feature classifiers for fast feature selection, *Advances in Knowledge Discovery and Management* **527**, pp. 113–131 (2014).

[41] H. Abdi and L. J. Williams, Principal component analysis, *WIREs Computational Statistics* **2**, 4, pp. 433–459 (2010).

[42] S. T. Roweis and L. K. Saul, Nonlinear dimensionality reduction by locally linear embedding, *Science* **290**, 5500, pp. 2323–2326 (2000).

[43] A. M. Martínez and A. C. Kak, PCA versus LDA, *IEEE Transactions on Pattern Analysis and Machine Intelligence* **23**, 2, pp. 228–233 (2001).

[44] J. Cordonnier, A. Loukas and M. Jaggi, On the relationship between self-attention and convolutional layers, in *ICLR, Procs.* (2020).

[45] N. A. Syed, S. Huan, L. Kah and K. Sung, Incremental learning with support vector machines, in *IJCAI workshop, Procs.*, pp. 496–505 (1999).

[46] A. Almaksour and É. Anquetil, Fast incremental learning strategy driven by confusion reject for online handwriting recognition, in *ICDAR, Procs.*, pp. 81–85 (2009).

Chapter 11

A Comprehensive Unconstrained, License Plate Database

Nicola Nobile[1], Hoi Kei Phoebe Chan[2] and Marleah Blom[3]

Concordia University, CENPARMI, Montreal, Canada
[1]nicola@cenparmi.concordia.ca, [2]hkpc07@gmail.com,
[3]marleah@encs.concordia.ca

Abstract. In this chapter, information about a large and diverse database of license plates from countries around the world is presented. CENPARMI's growing database contains images of isolated plates, including a small percentage of vanity plates, as well as landscapes, where the vehicle is included in the image. Many images contain multiple license plates, have complex scenery, have motion blur, and contain high light and shadow contrast. Photos were taken during different seasons, many are occluded by foreground objects, and some were taken through mist, fog, snow, and/or glass. In many cases, the license plate, camera, or both were in motion. In order to make training more robust, different cameras, lenses, focal lengths, shutter speeds, and ISOs (sensor sensitivity) were used. A summary of proposed guidelines used for license plate design is outlined, details about the database of license plates are presented, followed by information about ongoing work and efforts for labelling/ground truth.

Keywords: License plates, Database, Automated license plate readers

1 Introduction

License plates from around the world come in a wide range of formats, character sets, restrictions, fonts, and layouts. CENPARMI is currently building a large database of license plates from several countries. Although there are several license plate databases, they are limited as they contain plates that are limited to only one region, have few points-of-view of the plates, and were taken in ideal environments. We wanted to create a more challenging database that includes images of license plates (a) from many countries, states, and provinces, (b) taken from many different angles and heights, (c) are on stationary and moving vehicles, and (d) are located in environments that encounter differing weather conditions such as snow and rain.

1.1 Background Information

Many countries issuing license plates belong to the American Association of Motor Vehicle Administrators (AAMVA). The AAMVA creates standards for road safety and investigates topics related to the registration of vehicles. Members must ensure that license plates conform to legibility guidelines established by the association. License plates are stated to be more effective when they are designed to optimize legibility to the human eye as well as for Automated License Plate Readers (ALPRs) [1]. ALPR is also used when discussing Automatic License-Plate Recognition systems. The terms will be used interchangeably throughout this chapter. As a result, the AAMVA imposes strict standards to make identification easier for both humans and machines. Their relevant policies are related to display and design. Some additional AAMVA proposed guidelines for license plate design are shown in Table 1 [1].

Table 1. Some AAMVA Proposed Guidelines

1. Characters are at least 2.5 inches in height, proportionally wide, and spaced no less than 0.25 inches apart.
2. Character stroke weight (thickness of lines) are between 0.2 and 0.4 inches.
3. Characters are positioned on the plate no less than 1.25 inches away from the top and bottom edges of the license plate.
4. The font and spacing present each alphanumeric as a distinct and identifiable character. Standardized fonts and font sizes that clearly distinguish characters are used.
5. Non-alphanumeric characters are allowed. They are considered part of the license plate number and are to be accurately displayed on the license plate.
6. Graphics on license plates must not distort or interfere with the readability of the characters or with any other identifying information on the plate.
7. Graphics must be to the right or left side of the license plate number.
8. A background is allowed but must not interfere with the ability to read the license plate number by both a human and by an ALPR.

As indicated in the AAMVA's guideline number 4, fonts, font size and spacing much allow for each character to be easily identifiable. Many fonts can be found on plates as there are Asian characters (like traditional Chinese and Japanese), special characters (such as dashes and accents) and alphanumeric characters from North America and Europe. Some countries limit the character set based on human visual identification or more recently, based on results by computerized recognition systems and cameras. For example, the letter "I" and the number "1" as well as "B" and "8" look very similar from far away. In this case, the issuing authority may ban the letters "I" and "B" to improve both legibility and to avoid near duplicate plates.

Vanity plates must obey some restrictions, yet they do not necessarily need to follow the standard plate format of a given district. For example, in Québec, Canada, government issued license plates follow a "LDD LLL" or a "LLLDDDD" pattern [2]. Where "L" is a letter and "D" is a digit. This can cause additional issues for ALPR as the system can no longer take advantage of particular patterns. Vanity plates as long as "JIMIDAR" and "VITRIOL", and as short as "7Z", "MV", "HEY", and "6888" have been observed.

Car owners are required to maintain their plates in good condition and obey certain restrictions. Tinted covers are illegal in places like Ontario [3] and some states such as Connecticut in the United States, do not allow plates to be covered by frames [4]. Rules may not always be in place nor are they always adhered to due to dirt, snow, or other coverings, such as shields or frames. Information displayed on the license plates may be obstructed or may be difficult to see at night. For example, we have seen license plates so dirty that it was difficult to read the contents, especially snow-covered plates, plates with tinted or smoked bubble shields in front of them. Some license plates have a frame that partially covers the letters and/or digits of the plate. Also, there were some unlit plates, which makes reading them difficult at night.

2 Database Overview

Some issuing authorities may use decals, including stickers, to add graphics to license plates. These are for presentation purposes only but can influence identification. To renew a license plate, some places still use stickers to indicate the expiration date. Additionally, governing agencies may allow for a background image, logos or icons to be displayed. Veteran plates are known to include background graphics and logos, all of which greatly affects character segmentation and recognition. Finally, license plates may be different colours. For example, electric cars in Québec have license plates in a light green colour, which may affect plate identification if a colour threshold is being used.

Our goal is to create a large vehicle license plate database. It was decided to collect a variety of images of license plates from around the world to make ALPR systems more robust. Users can then choose to focus their recognizers on one specific plate type if so desired.

Other datasets of license plates exist, including the Reid dataset [5]. It was intentionally made of low-quality images, extracted from 9.5 hours of video. This dataset simulates surveillance cameras by placing the Full-HD video capture devices on bridges above eight highway locations. The Reid dataset contains 76, 412 color license plate images of different lengths, as well as blurred images and images that included slight obstructions. Having extracted plates taken from the same viewpoint was perceived to be a limitation. Many surveillance systems capture vehicle images close to the ground, including police cars mounted with ALPR systems, Closed-circuit television (CCTV) cameras that monitor parking lots, and CCTV cameras around London, England. It was thus deemed important to create a dataset with license plate images from different heights and distances.

The high dynamic range (HDR) dataset [5] was captured by digital single-lens reflex (DSLR) cameras with three different exposures. The limitation with this particular dataset is that the vehicle needs to be stationary in order to dynamically merge the three different exposures of the same scene. Therefore, identification of plates on moving vehicles on highways and streets is limited. Our license plate dataset contains a broader range of situations by including cases where the vehicle and camera are stationary, moving, or a combination of the two.

The CENPARMI License Plate Database currently consists of 3,322 images of license plates from seven countries. Table 2 shows the distribution of the current state of our database. We currently have 136 unclassified plates partly because vehicles were from out-of-town, and partly because of nonstandard AAMVA plates, such as military issued plates.

Table 2. License Database Country Distribution

Country	Count
China	151
Japan	13
Canada	1,806
U.S.A.	211
Jordan	199
Palestine	302
Turkey	468
Unclassified	172
Total:	3,322

Some existing images of plates within the CENPARMI database are clean, legible, and taken at a close distance. Meanwhile, some plates are challenging to read. Some plates are blurry due to photographs being taken 1) from a long distance, 2) while cars are moving, 3) from various angles, 4) while walking, and 5) under challenging weather conditions (such as rain, slush, and snow). Plates are also blurry as they 1) were broken, bent, or rusted, 2) have colorful backgrounds, 3) have stickers or other markings, 4) have difficult fonts, 5) have varied font colors, 6) have differing compositions, 7) are protected by a tinted or smoked cover, and 8) include frames from dealership or other organizations. **Fig. 1** includes samples, taken from our database, that challenge plate recognition such as vanity plates (a, c, e, g), plates with logos (c, e), plates with background images (b, h), plates with obstructing frames (b, f), light coloured plate (g), plates with stickers (b, h, i) and plate with smoked bubble shield (d).

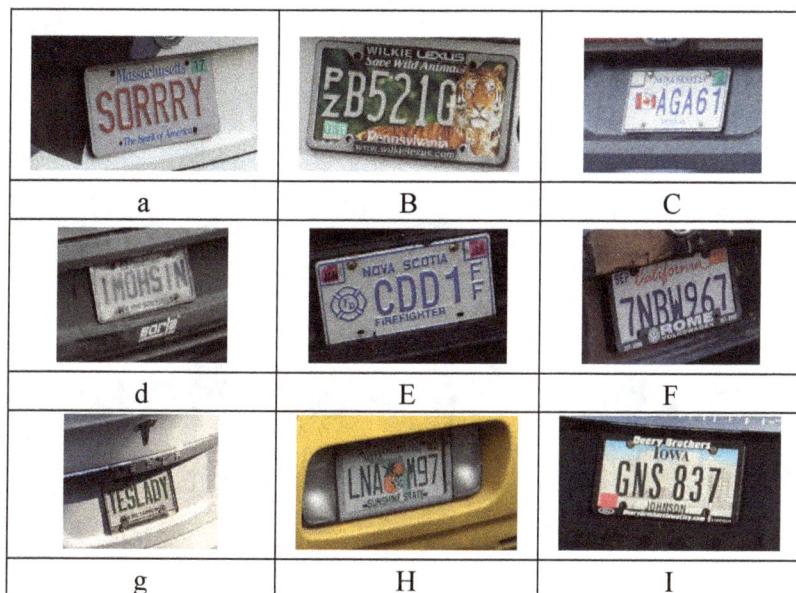

a	B	C
d	E	F
g	H	I

Fig. 1. Samples of License Plates that Challenge Plate Recognition

Our database is unconstrained, which can make it ideal for research. For example, Silva and Jung [7] identify vehicles from an unconstrained scene, locate the license plate, rectify the plate, and then perform OCR on the corrected plate image. Images were taken from several vantage points. The types of images included, and method used to take photographs, thus make the dataset a valuable contribution.

3   Data Collection

Images were taken at differing distances, including macro scales (isolated plates) and landscape views, whereby other objects and/or vehicles appear in the photo. Similar to Luo et al. [6], who performed research on detecting multiple license plates in complex scenes, many images within the database contain several vehicles and license plates.

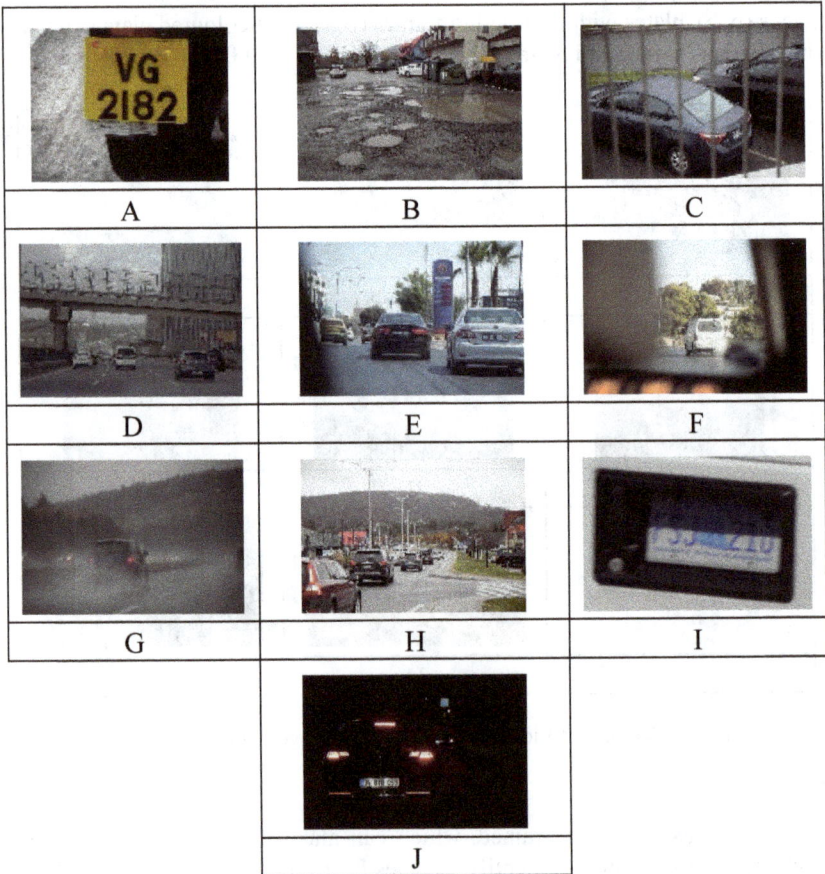

Fig. 2. Database Photo Variations

　　To maintain privacy, we took special care to remove identifiable people from photos by either cropping, blurring, or using Photoshop to remove them. **Fig. 2** shows some variations in our database in terms of distance and angle of the plate, the surrounding conditions, and the camera placement in relation to the plates. The images in **Fig. 2** include license plates that are close and isolated (a, i), far from the camera (b, d, f), occluded (c, f), both vehicles and camera in motion (d, e, f, g), contain multiple license plates (d, e, h), mist conditions (g), shot through a car window (d, e, f, g), vehicles in motion (d, e, f, g, h, j), still camera (a, b, c, i, j), high vantage point (c), partly shaded (i), and taken at night (j).

To prevent overfitting, we used several camera models, lenses, and resolutions. Both high-end professional Single Lens Reflex (SLR) cameras and wireless phone cameras were used. Table 3 lists the cameras used. Handheld cameras were used to capture most images. As a result, some images may be rotated by a few degrees. Both landscape and portrait orientations are represented.

Table 3. License Plate Capture Devices

Camera Maker	Camera Model	Camera Type
Canon	EOS REBEL T2i	SLR
Canon	EOS 6D Mark II	SLR
Sony	DSC-WX220	Point and shoot
Apple	iPhone 6	Phone
Apple	iPhone 6s	Phone
LGE	Nexus 5X	Phone
Samsung	SGH-I337M	Phone
Sony	H8324	Phone
Vivo	X23 Magic color	Phone

Resolutions vary based on the capture device. SLR cameras, with high end lenses, provide a higher resolution and photo size. Most of the time, the white balance was set to automatic. This is mostly the case for the phone cameras. For the SLR cameras, sunny, cloudy, or a manual white balance may have been chosen for some photos. The lenses used range from wide angle to telephoto lenses.

4   Ongoing Work

As we are in the early stages of collecting images of license plates, we plan to move forward by using different capture devices and lenses as well as increasing the number of samples and types of images. It has been harder to add new plates to the database due to COVID-19 based travel restrictions imposed throughout Canada. The Canadian border is closed for non-essential travel in and out of the country to control the pandemic [8]. Many provinces and territories also discourage interprovincial travel by imposing a 14-day mandatory quarantine and self-isolation period or promote a 'travel bubble' like the 4 Atlantic provinces [9]. These restrictions thus impact the types of plates that are available. Therefore, we are currently focused on capturing local or foreign plates found on vehicles allowed to travel in Canada. We are looking to collaborate with people from various locations to send images of license plates to us.

Our database will not be limited to standard and outdated images. It will grow to include current and future plates. For example, more regions are allowing for more variations in their plates, including color changes, background images, vanity plate options, change in typesetting, inclusion of logos, etc. For example, drivers in Arizona recently were given five new specialty license plate options [10], now having a total of 71 choices for that state. The province of Ontario also recently issued a new blue license plate [11] with white lettering, effectively inverting the current color scheme for that province. However, the government of Ontario has reversed its decision to add these new plates [12]. Personalized plates with emojis are now available to vehicle owners in Australia [13]. In addition, some states are experimenting with digital license plates [14] that makes the physical change from a metal plate to a Liquid Crystal Display (LCD) or e-Ink display. Participants of digital plates have the choice of using the standard white background with black lettering, or the inverted black background with white lettering. Static license plate databases may not be ideal to train and test a dynamic ALPR system and thus we will continue to work to make this set robust and relevant.

4.1 Labeling/Ground Truth

We manually extracted some isolated characters from a few selected license plates. However, because we intend to drastically increase the number of samples, this will not be a feasible method. We are in the process of building a semi-automatic tool to assist in this labeling procedure. The software will locate the license plate(s) within an image, perform OCR on the plate and attempt to identify it, and extract the isolated characters from them. The user will have the ability to change the automatic proposed labels.

Information we intend to include for each image will contain the capture device (camera make and model), resolution, lens, focal length, number of license plates in the image, the location of the plate(s) within the image, the content of the plates including any special symbols, and the country the plate was issued. Temporal information such as time of day and season, and physical information such as plate occlusion and the weather situation such as rain, snow, mist, or fog, that can affect recognition, will be included as well.

5   Conclusion

License plate databases are important not only for researchers, but also for commercial Automatic Number-Plate Recognition (ANPR) and ALPR systems. These systems need to be trained on the latest license plate formats. In addition, police and law enforcement around the world can use ALPR to check if a vehicle is registered, licensed, or to track a lost or stolen vehicle using their car-mounted cameras. Some electronic toll booths collect payments from drivers based on the license plate image of the vehicle as it passes through the gate, thereby speeding up the collection process. Local or state governments may use images to monitor the movement of traffic. This information can be used for urban and social planning purposes as it allows governments and agencies to determine whether more roads or highways need to be built or how traffic is affecting health and safety of people living and travelling in the area. The CENPARMI comprehensive and unconstrained database thus, even in its initial stages, provides a novel collection of images of license plates, which contributes to research as well as commercial and security related initiatives.

References

[1] American Association of Motor Vehicle Administrators (AAMVA), "License Plate Standard," 2016. [Online]. Available: http://www.aamva.org/LicensePlate Standard_2016. [Accessed 09/02/ 2021].

[2] Société de l'assurance automobile du Québec (SAAQ), "Categories of Licence Plates," 2020. [Online]. Available: https://saaq.gouv.qc.ca/en/vehicle-registration/categories-licence-plates/. [Accessed 15 September 2020].

[3] CityNews, "Licence Plate Covers Not Permitted," 17 July 2006. [Online]. Available: https://toronto.citynews.ca/2006/07/17/licence-plate-covers-not-permitted/.

[4] EyeWitness News, "Local woman warning drivers of license plate frame law after receiving ticket," WFSB (Meredith Corporation), 7 March 2019. [Online]. Available: http://www.wfsb.com/news/local-woman-warning-drivers-of-license-plate-frame-law-after/article_39ec11ba-414b-11e9-8eef-ef030a73f602.amp.html.

[5] J. Španhel, J. Sochor, R. Juránek, A. Herout, L. Maršík and P. Zemčík, "Holistic Recognition of Low Quality License Plates By CNN Using Track Annotated Data," in *2017 14th IEEE International Conference on Advanced Video and Signal Based Surveillance (AVSS)*, Lecce, 2017.

[6] Y. Luo, Y. Li, S. Huang and F. Han, "Multiple Chinese Vehicle License Plate Localization in Complex Scenes," in *2018 IEEE 3rd International Conference on Image, Vision and Computing (ICIVC)*, Chongqing, China, 2018.

[7] S. M. Silva and C. R. Jung, "License Plate Detection and Recognition in Unconstrained Scenarios," in *2018 European Conference on Computer Vision (ECCV)*, Munich, Germany, 2018.

[8] Government of Canada, "Coronavirus disease (COVID-19): Travel restrictions, exemptions and advice," 10 September 2020. [Online]. Available: https://www.canada.ca/en/public-health/services/diseases/2019-novel-coronavirus-infection/latest-travel-health-advice.html. [Accessed 15 September 2020].

[9] M. Collie, "Can I go to another province? The latest coronavirus travel restrictions, by region," Global News, Corus Entertainment Inc., 25 June 2020. [Online]. Available: https://globalnews.ca/news/7106284/coronavirus-provincial-travel/. [Accessed 15 September 2020].

[10] R. Randazzo, "New for your car: Arizonans have 5 new specialty license plate options, creating 71 total choices," The Arizona Republic, 10 August 2020. [Online]. Available: https://www.azcentral.com/story/money/business/consumers/2020/08/10/arizonans-have-5-new-specialty-license-plate-options/3335954001/. [Accessed 15 September 2020].

[11] CBC News, "Ontario's new blue licence plates hitting the road," 8 February 2020. [Online]. Available: https://www.cbc.ca/news/canada/ottawa/new-blue-ontario-licence-plates-1.5454205#:~:text=1%2C%20all%20new%20plates%20issued, letters%20instead%20of%20the%20opposite. [Accessed 15 September 2020].

[12] K. Goodfield, "Ontario will not be moving forward with faulty blue licence plates," CTV News Toronto, 06 05 2020. [Online]. Available: https://toronto.ctvnews.ca/ontario-will-not-be-moving-forward-with-faulty-blue-licence-plates-1.4927535. [Accessed 23 09 2020].

[13] Queensland Government, "Personalized Plates Queensland," Queensland Government, [Online]. Available: https://www.ppq.com.au/_/media/files/pdf/downloadable-ebrochure-2020v2.pdf?la=en. [Accessed 15 September 2020].

[14] State of California Department of Motor Vehicles, "Report On Alternative Registration Products Pilot Program," 2019. [Online]. Available: https://www.dmv.ca.gov/portal/uploads/2020/04/AlternativeRegistrationProducts.pdf. [Accessed 09 02 2021].

Chapter 12

Classification of Spanish Criminal News Using Neural Networks

Mireya Tovar Vidal and Emmanuel Santos Rodríguez

Benemérita Universidad Autónoma de Puebla, Facultad de Ciencias de la Computación, Puebla, México
mtovarvidal@gmail.com, e.ss.rdz@gmail.com

José A. Reyes-Ortiz*

Universidad Autónoma Metropolitana, Unidad Azcapotzalco, Departamento de Sistemas, Ciudad de México, México
**corresponding: jaro@azc.uam.mx*

Abstract: With the rapid expansion of digital newspapers, readers have an overwhelming amount of news available daily. However, it is difficult to keep track of the news that is only of interest to the reader or filtering those that belong to a specific topic or crime, e.g. assault or kidnapping. The main challenge is handling free texts from Spanish criminal news; however, deep learning has proposed models of representation and manipulation of natural language that help in this problem. Therefore, this chapter proposes the use of deep neural models to classify criminal news published by Mexican digital newspapers. The whole approach includes a word embedding model for representing the documents and several layers in neural networks to classify news text. We experiment with three deep neural models: Bidirectional LSTM, Attention Bidirectional LSTM and BERT-based model. According to the experimental results, our system shows promising results.

Keywords: Natural Language Processing, News classification, Neural Networks, Deep Learning, Spanish Criminal News

1   Introduction

Currently, we have a large amount of data, which increases every second and this trend will only increase in the future. Most of them are unstructured data, e.g. news, social messages, scientific texts or web texts. They are available through Internet, but also in various media as social networks. However, it is difficult to analyse such magnitude of information manually due to the lack of a structured language model that allows it to be handled automatically.

News provides fresh and updated information about events or facts. An analysis of them involves a tracking or filtering of specific news that are of interest to a reader, for example those that belongs to crime (killing, assault, kidnapping, rape). News headlines are available from social networks like Twitter and they are a powerful resource for extracting knowledge about certain domain.

The free text needs to be processing and the Natural Language Processing discipline is a great help. It proposes a set of tools to exploit automatically the knowledge contained in text. Also, Text classification, the task of assigning labels to a text according to its content, is a helpful to analyse texts and find non-explicit knowledge from them. Such task allows us to apply text classification in different areas such as spam detection, sentiment analysis, content detection inappropriate, social network monitoring, among others [1]. However, there are different approaches to perform this task as rule-based systems, machine learning based, deep learning or hybrid systems.

In this chapter, an approach to classify Spanish crime news based on the headlines from Twitter using deep neural networks is proposed. A baseline is provided to compare the performance of three neural models in terms of *accuracy, precision, recall* and F_1. The whole approach involves the following steps: a) as starting point, the approach considers headlines of criminal news published by Mexican digital newspapers; b) a word embedding model for representing the documents as a dense vector with real values; c) three neural networks are implemented to classify criminal news, such as: Bidirectional LSTM [2,3], Attention Bidirectional LSTM [4] and BERT-based model [5]; d) an evaluation process was carried out using a gold standard dataset, which was extracted from twitter based on hashtags (words accompanied by the # character) and validated by human; e) the three neural models are evaluated and compared to each other in order to determine the best model for the given task. The classification task consists of assigning one of the class labels (rape, kidnapping, homicide, suicide, assault or exploitation) to each Spanish criminal news as untagged input text. According to the experimental results, the best performance was achieved by BERT-based model reaching 98.65 % precision.

The rest of this document is structured as follows. Section 2 presents a description of works related to news classification through different approaches. In Section 3 the solution approach is presented introducing the details of each process and the three neural network models. Section 4 discusses the obtained results for the classification of Spanish criminal news using three neural networks and comparing them. Finally, Section 5 presents the conclusions and future work.

2   Related work

There are different proposals for text classification, especially for news classification. However, most of these are focused in English news. Therefore, below, various solutions are shown for the problem mentioned above.

[6] proposes a comparison of three machine learning techniques (ANN, MultiBinomial Naive Bayes and kNN) to tackle Spanish news classification from La Capital newspaper. The proposed approach relies on $TF - IDF$ and dimensionality reduction to weight the documents approximately equal. The best technique for this corpus is MultiBinomial Naive Bayes, however they didn't perform extensive tests, nor did they provide a benchmark.

The system presented in [7] consists in a three-phase system (preprocessing, learning and classification) built upon a Support Vector Machine for classifying Indonesian news. There are four categories for the news the classification is approached as a one against one (pairwise classification). This system achieves a good accuracy (85%), however the authors suggest trying another multi-class mechanism or diversifying news sources.

[8] describes a news classification system for an Indian language, Punjabi, using Naive Bayes classifier. This system classifies the news into four categories: terror attack news, murder related news, accidental news and suicide news. The news set is obtained from different news websites written in Punjabi. The performance of this proposal is measured through precision, recall and F_1 and it achieves a reasonable success even though Punjabi has a poor resource pool for preprocessing tools.

In [9] a system is proposed for classifying online news based on Hidden Markov Model for feature extraction and multiclass Support Vector Machines to perform classification. A comparison shows that this approach outperforms kNN and SVM in terms of accuracy. Nevertheless, there is ambiguity in the features that may lead to misclassification. It is important to highlight that the preprocessing phase for this system is crucial because it reduces computing time and noise.

[10] introduces a method to classify news as from the headlines shared on Twitter by Sri Lankan news groups. To extract the features, they used Bag-of-Words approach, but they reduced data dimensionality by removing the most and the least frequent words, as it's likely that these don't carry important information. For the classification task, a Support Vector Machine is used as it supports high dimensional data. The system performance is measured through effectiveness which is defined as the system ability to satisfy the user in terms of the relevance of the retrieved messages. At the same time, the effectiveness is measured by precision and recall. The overall effectiveness of the proposed system is good, except for one of the classes.

In [11], it is demonstrated how the text is classified using RNN variants such as LSTM, Bi LSTM and GRU. The experiments are performed on three datasets. We have special interest in the amazon books review dataset because the authors only used review titles to perform sentiment analysis. Also, they compare this approach to a Bag-Of-Words one. The provided results show that the all neural networks models, especially Bi LSTM, outperform the Bag-Of-Words approach in terms of accuracy by a considerable margin. However, it should be considered using full reviews on the Amazon dataset.

[12] discusses the high dimensionality problem associated created by selecting irrelevant features for classification, resulting in a declining accuracy. They address this problem using information gain for feature reduction in a first phase and Maximal Marginal Relevance for Feature Selection to select features in a second phase. The classification in done using Naive Bayes classifier. This two-phase system is proven to achieve good accuracy with reduced computational cost.

In [13], a system is proposed, which classifies news into sections for each newspaper. The news used were gathered in a span of six months from twitter accounts of three Mexican newspapers. This approach relies in a vector representation of the news articles content to subsequently apply different classifiers such as decision trees, Naive Bayes, SVM and logistic regression. The results show that the best classifiers for this task are SVM and logistic regression.

In [14], they use different machine learning models to label news articles into twelve categories, i.e. business, politics or technology. They compare the performance obtained by Google's BERT with more traditional methods such as SVM, Naive Bayes and Random Forest. According to the authors, BERT outperforms the other used approaches both in balanced and imbalanced datasets.

[15] proposes a three-step method named BERT-LB, which combines BERT and LSTM, to predict Chinese news topic. BERT-LB is shown to outperform BERT alone and other models used in Chinese text classification tasks, in terms of accuracy and F1. However, the authors theorise that if the news corpus is short, the proposed model may not be able to be used to its full potential and its performance will be closer to BERT alone.

[16] presents a deep learning model based on an attention mechanism to classify academic news. The approach is proven to achieve better performance in two different datasets, compared to traditional classification methods such as logistic regression, SVM or deep learning methods such as CNN or RCNN. However, it is important to mention that this work is mainly focused on Chinese text.

The work presented in [17] shows a comparison between a deep learning model (Bidirectional LSTM-RNN) and traditional approaches for text classification. The comparison, which was performed on THUC News dataset, demonstrated that the deep learning method obtains better accuracy rate and *F1* score. Furthermore, the authors state that they proposal conserves more contextual information and enables the semantic of the text to be expressed accurately, which in turn, results in better generalisation.

Criminal events are characterised in [18] by using ontology model to represent victims, offender, causes, time and location. The author presents an approach based on linguistic patterns to extract criminal events and populate an ontology, based on a previous works [19]. In this manner, the general event detection or extraction task using social media text is related to classification of criminal news. Several approaches to accomplish such task has been proposed. A machine learning approach for extraction new categories of events from Twitter is presented in [20]; semantic patterns are used by [21] to extract global event from English news; A event classification method based on inference rules is exposed in [22] in order to extract security-related events from online news.

In recent years, deep learning approaches have become popular and text classification centred on events is not been neglected. In [23], a deep neural network with word embedding is used to extract Chinese events form newswire documents; [24] propose a convolutional neural network for event-driven stock market prediction from news text, using both short-term and long-term influences of events on stock price movements; a deep neural network to extract social events from Twitter is presented in [25], which learns, as well as conducts, detection and extraction by using a bidirectional long short-term memory (LSTM) as a representation layer.

3   Proposed approach

This section presents three models of deep neural networks. First the Bidirectional LSTM and Attention LSTM models are presented with some fusion due to the similarity between their layers. Later, BERT-based model is described in detail.

Both Bidirectional LSTM (Bi-LSTM) and Attention LSTM (A-LSTM) share the following layers: input, embedding, LSTM, dropout and output. In the case of Bi-LSTM, the specific layers are max pooling and fully connected. In A-LSTM model, an attention mechanism layer is considered. They are described below.

3.1   *Word embeddings*

The input layer for this model is a dense vector representation of the words in the document fed into the input, where each word is represented by a real value. This representation, called word embeddings, provides some advantages over sparse word representations, such as One-Hot Encoding, because it requires less dimensions and it preserves better the semantic of the words. Given the relatively small dataset, we decided to use pre-trained embeddings rather than training the word embeddings. Therefore, we employ Spanish word embeddings, which are trained on the Spanish Billion Word Corpus [26] using Word2Vec.

3.2   *Long Short-Term Memory*

Recurrent Neural Networks (RNN) can use their memory to process input sequences, thus they can deal with sequential data such as text. However, they suffer from vanishing or exploding gradient problems [27]. To overcome this problem, a new architecture was designed: Long Short-Term Memory (LSTM) which can effectively deal with long term dependencies through its gate mechanism. LSTM units are mainly composed of three elements: input gate, output gate and forget gate. The input gate controls the information added to the cells, the forget gate decides what information should be kept or removed and the output gate decides what parts of the cell state will be output. However, an in-depth look of this architecture can be found in [2, 3].

For this approach, we want to exploit both future and past information, therefore we propose a bidirectional Long Short-Term Memory. As stated in [28], this architecture is more effective than unidirectional LSTMs and is appropriate for tasks where context is crucial. Also, it's convenient to have access to future and past context for sequence modeling tasks [4].

3.3   *Regularisation*

Dropout [29] is a regularisation technique where the core idea is to randomly drop units as well as their connections during training to prevent over-fitting, improving the network performance in a variety of domains. We use dropout in the penultimate layer of the model.

On the other hand, we apply L2 Regularisation for the LSTM layers, which is defined as shown in Eq. 1.

$$|W|_2^2 = \sum_{i=1}^{m} |w_i|^2 \tag{1}$$

3.4   *Max pooling*

We employ a one-dimension global max pooling layer after the LSTM layer to extract representative information. This reduces the dimensionality but keeps the key information. However, this operation loses information about locality of the words in a text. In [30], the impact of the pooling strategy is discussed, and it is stated that 1-max pooling outperforms other pooling strategies for sentence classification. This layer is characteristic of Bidirectional LSTM model, as presented in Fig. 1.

3.5   *Attention mechanism*

Attention allows recurrent neural networks to assign more attention to specific parts of the input that be important as mention in [31], who define Attention as shown in Eq. 2.

$$c^{<t>} = \sum_{t'} \alpha^{<t,t'>} \alpha^{<t'>} \tag{2}$$

$\alpha^{<t,t'>}$ notes the amount of attention that the output $y^{<t>}$ should pay to activation $\alpha^{<t'>}$ and context $c^{<t>}$ at time t. The attention weight is the amount of attention that the output $y^{<t>}$ should assign to activation $\alpha^{<t'>}$ is given by $\alpha^{<t,t'>}$ and is computed as shown in Eq. 3.

$$\alpha^{<t,t'>} = \frac{exp(e^{<t,t'>})}{\sum_{t''=1}^{T_x} exp(e^{<t,t''>})} \tag{3}$$

Since its introduction, attention has become a valuable mechanism in a wide variety of Natural Language Processing related problems. It offers

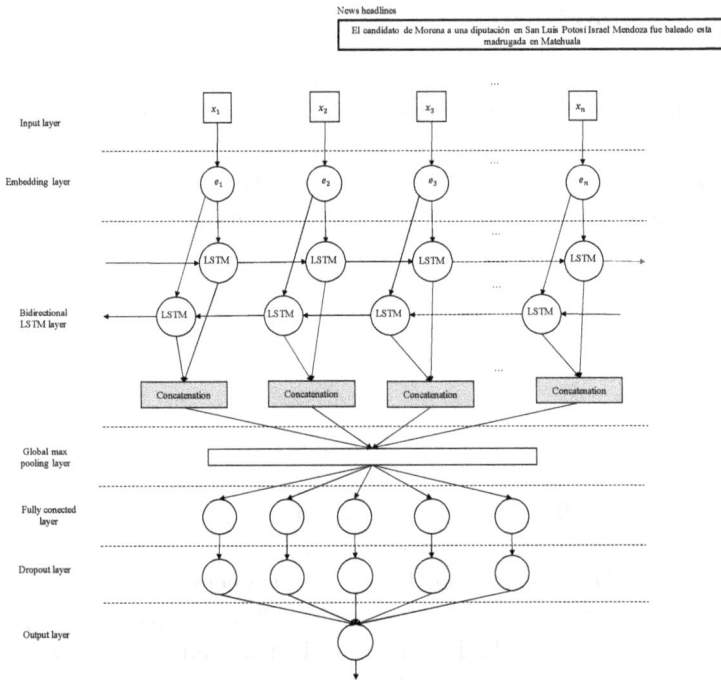

Fig. 1   Overview of the Bidirectional LSTM model

advantages such as performance gains and provides tools to get a better and more readable understanding of neural networks outputs [32].

In contrast to Bidirectional LSTM, we use a many-to-one attention mechanism to replace the global max pooling and fully connected layers, but the rest of the model remains the same (Fig. 2).

3.6   *BERT-based model*

The BERT-based model is considered a deep neural networks and it used a pre-training Bidirectional Transformers for language understanding [5]. A transformer architecture based on BERT is applied to classify unlabeled text using transfer learning. BERT starts with the use of a *[CLS]* special symbol added in front of every input and the text separated by tokens; words embeddings are generated by considering the sum of the token embeddings, the segmentation embeddings and the position embeddings. The pre-trained BERT-based model is used over words embeddings to generate the output layer that is depicted at the token level. This output layer includes a class label (C) in the *[CLS]* position for each input example, which is used for several classification task.

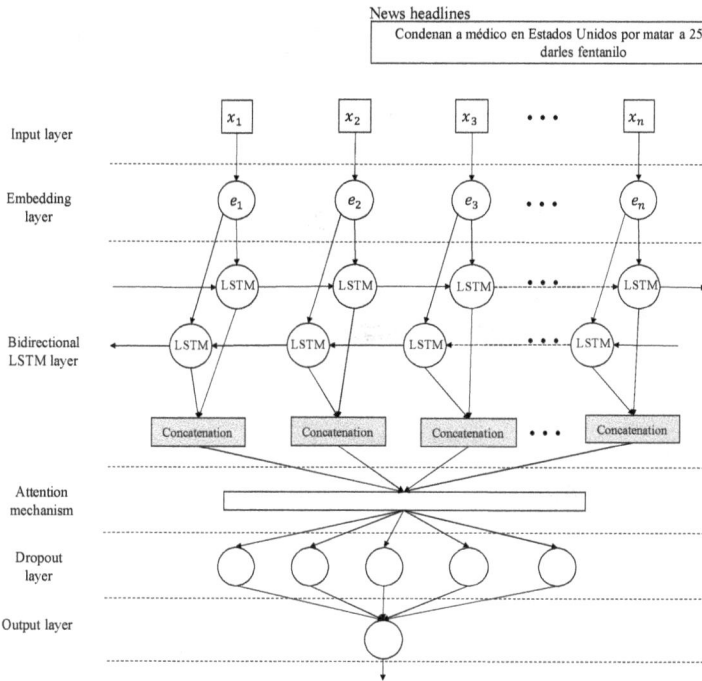

Fig. 2   Overview of the Attention LSTM model

The tokens of each Spanish criminal news are depicted over the layers in the BERT-based deep neural model, leaning on the pre-trained model, in order to determine the corresponding class label to each criminal news.

The Fig. 3 shows the BERT-base model including a criminal news as an input example. The possible values for the class label of the output layer are: rape, kidnapping, homicide, suicide, assault or exploitation.

4   Experimental results

In this section, a description of the used datasets is analysed, and the results obtained are detailed and explained.

4.1   *Dataset description*

In this work, two dataset are used: a) the first datasets, which is provided in [19], consists in news headlines from 24 Mexican news groups Twitter accounts collected from November 2017 to June 2019 for a total of 7988 tweets, focusing only in crime news; b) the second dataset contains 2295 Spanish criminal news, which was collected between June 2019 and April

News headline

| "Condenan a médico en Estados Unidos por matar a 25 pacientes al darles fentanilo" |

Input sentence — [CLS]   condenar   médico   • • •   fentalino

Embedding layer — e_{CLS} e_1 e_2 • • • e_n

BERT-based

Token level output layer — c t_1 t_2 t_n

Class label

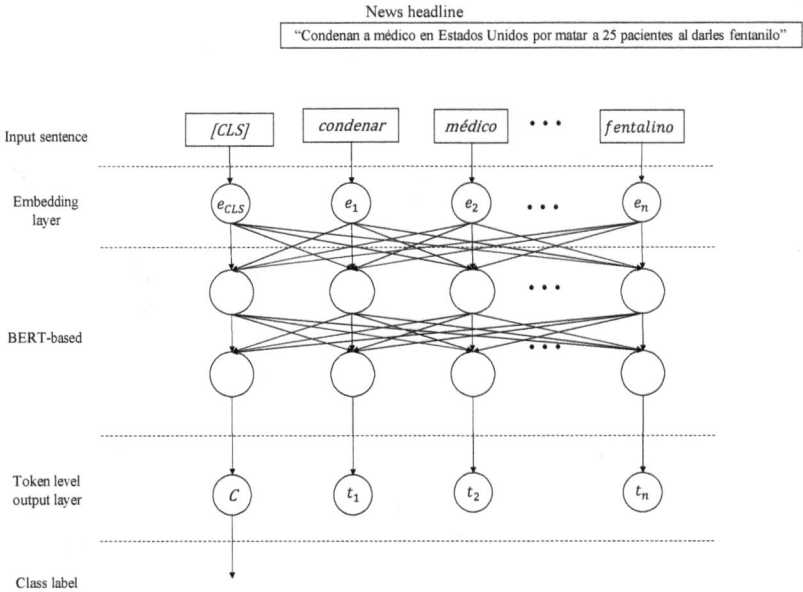

Fig. 3   Overview of the BERT-based model

2020 from Mexican Twitter accounts. The criminal news was extracted using hashtags from Twitter such as: #asalto, #secuestro, #abuso, #asesinato, #explotación, #violación.

Each tweet is assigned a label out of six (rape, kidnapping, homicide, suicide, assault or exploitation). Then, humans validate each news and its label assignment to compose gold standard datasets. It is important to highlight that the classes are imbalanced since most crimes committed are homicides and assault. As an example, Fig. 4 shows the sources for the news in the first dataset and it is observed that *Noticieros Televisa* is the account from which most tweets percentage were obtained (1639 tweets) while the account of *Alejandro Marti* represents the least percentage (just a tweet).

Both datasets were split in 60% for training, 20% for validation and 20% for test.

4.2   *Baseline*

Support Vector Machine is placed as baseline for classifying the news. First, we applied preprocessing (removing punctuation, stop words and converting the sentences to lowercase) to the news headlines, then we

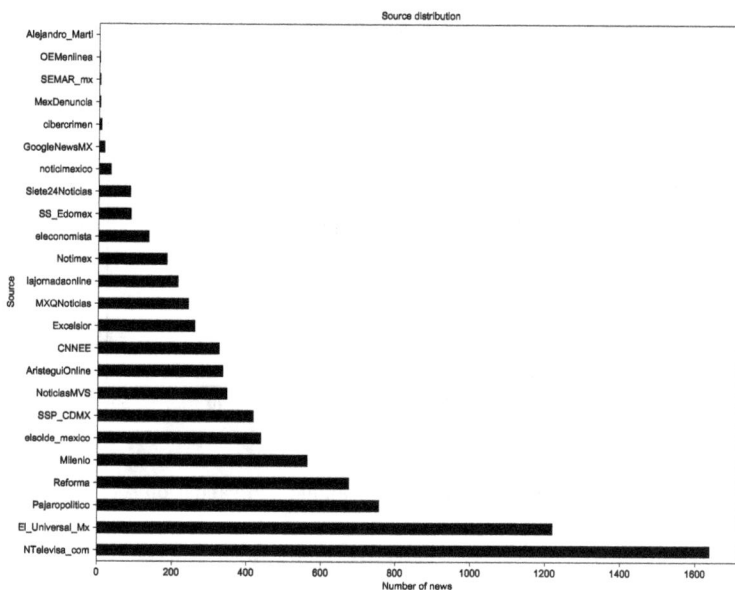

Fig. 4   Source distribution for the dataset

converted the headlines into vectors using a One-Hot encoding with 10,000 features.

4.3   *Bidirectional LSTM*

In order to implement Bi-LSTM model, we used Keras with TensorFlow 2.0 as backend, which in turn allowed us to train the network using a GPU. The experiments were run on a GeForce GTX 1050Ti GPU and an Intel I5 CPU. The network is trained to minimise the binary cross-entropy loss, using ADAM optimisation with batch size 64 during 50 epochs. We performed Random Search [34] using Keras tuner to determine the number of units for hidden layers and the dropout rate. We used 190 units for LSTM and fully connected layers. The L2 regularisation used is on the order of 10-3 and the dropout was set to 0.15. For the embedding layer, we employ the 300-dimensional word vectors trained by Cardellino [26].

In Fig. 5 we can observe the plot of the model history for training and validation sets. According to the figure, it might be better to stop the training before epoch 25 using a callback function.

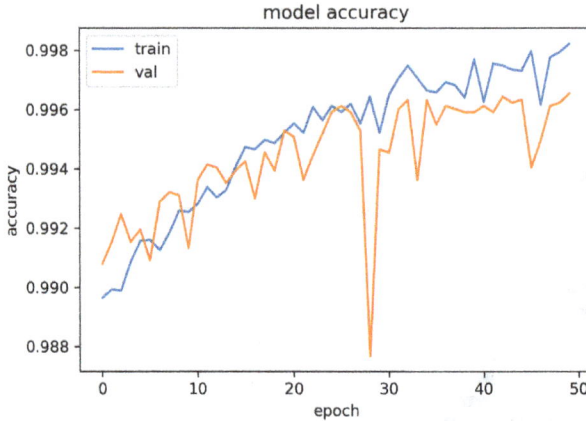

Fig. 5   Plot of accuracy for training and validation sets

4.4   *Attention bidirectional LSTM*

This network is trained to minimise the binary cross-entropy loss, using ADAM optimisation with batch size 32 during 60 epochs. We also performed Random Search to determine the number of units for the LSTM hidden layers and the dropout rate. Thus, we used 180 units for LSTM and the dropout was set to 0.25. The L2 regularisation used is on the order of 10-3. For the embedding layer, we employ the 300-dimensional word vectors as well as in the bidirectional LSTM.

4.5   *BERT*

The implementation of BERT-based model followed, for the training phase, the one cycle policy [35] with a learning rate of 2e-5 during four epochs; the feature number and maximum length of sequences were set to 10000 and 200, respectively. This implementation utilises k-train [36], a lightweight wrapper library for *Keras*, which allowed us to reduce development time.

4.6   *Results*

The results obtained by the proposed model in order to classify Spanish criminal news and the baseline were measured using Scikit-Learn metrics. We employ accuracy, precision and balanced accuracy as well as classification report to get a detailed insight.

After training our model for 50 epochs, we obtained the classifications results which are shown in Table 1. According to Table 1, it is observed

Table 1   Classification report for the Bi-LSTM model for the first dataset

Class	Precision	Recall	F_1	Support
Assault	99	100	100	542
Exploitation	100	100	100	26
Homicide	99	99	99	768
Kidnapping	97	97	97	76
Suicide	92	92	92	63
Rape	98	99	99	123

Table 2   Classification report for the Attention-LSTM model for the first dataset

Class	Precision	Recall	F_1	Suport
Assault	99	99	99	542
Exploitation	100	100	100	26
Homicide	98	99	99	768
Kidnapping	99	96	97	76
Suicide	90	90	90	63
Rape	100	98	99	123

Table 3   Classification report for the BERT model for the first dataset

Class	Precision	Recall	F_1	Support
Assault	99	100	99	542
Exploitation	100	100	100	26
Homicide	100	99	99	768
Kidnapping	97	99	98	76
Suicide	97	95	96	63
Rape	99	99	99	123

that the class with lowest scores is class *Suicide*, even though it has more support than class *Exploitation*.

The attention-LSTM (Table 2) model obtains similar results to the Bi-LSTM but slightly worse overall, except for Exploitation class where the results are the same.

In Table 3, we can observe the BERT model obtained better scores for the second class with the least support compared to the LSTM based models.

With respect to the baseline, the scores for precision are similar in Kidnapping and Rape classes, but for the rest are lower than the proposed model. This can be observed specially in the Exploitation class where our model scores 100% precision while the baseline obtains 60% (see Table 4).

Table 4   Classification report for the baseline model for the first dataset

Class	Precision	Recall	F_1	Support
Assault	93	92	92	542
Exploitation	60	23	33	26
Homicide	86	95	91	768
Kidnapping	99	91	95	76
Suicide	83	40	54	63
Rape	96	82	89	123

Table 5   Comparison of the baseline and proposed model using accuracy, precision and balanced precision model for the first dataset

Model	Accuracy	Precision	Balanced precision
Bi-LSTM	98.87	97.86	97.74
Baseline	89.54	86.23	89.54
Attention-LSTM	98.43	97.77	97.01
BERT	99.18	98.65	98.66

Table 6   Classification report for the Bi-LSTM model for the second dataset

Class	Precision	Recall	F_1	Support
Assault	99	100	99	701
Exploitation	100	77	87	26
Homicide	99	98	98	1275
Kidnapping	81	97	88	93
Suicide	88	88	88	66
Rape	97	98	98	133

However, the baseline recall and F_1 scores are worse than the proposal for the classes with the lowest support.

Nevertheless, when comparing the models, it can be seen in Table 5 that the deep learning approaches outperform the baseline in all the metrics used. But BERT is the best model overall by a small margin.

For the second dataset, the Bi-LSTM performs worse, especially in two classes (kidnapping and suicide) when compared to the other neural network models as shown in Tables 6, 7, and 8.

According to Table 9, the Attention-LSTM and BERT models are better for this dataset in terms of precision and balanced precision, in contrast to the first dataset where Bi-LSTM achieved almost equal results for all metrics.

Table 7   Classification report for the Attention-L-STM model for the second dataset

Class	Precision	Recall	F_1	Support
Assault	100	99	99	701
Exploitation	100	88	94	26
Homicide	98	99	99	1275
Kidnapping	97	94	95	93
Suicide	92	89	91	66
Rape	99	98	98	133

Table 8   Classification report for the BERT model for the second dataset

Class	Precision	Recall	F_1	Support
Assault	98	99	99	701
Exploitation	100	81	89	26
Homicide	99	99	99	1275
Kidnapping	97	97	97	93
Suicide	97	91	94	66
Rape	98	99	99	133

Table 9   Comparison of the baseline and proposed model using accuracy, precision and balanced precision model for the second dataset

Model	Accuracy	Precision	Balanced precision
Bi-LSTM	97.77	94.04	92.89
Baseline	97.82	97.82	94.93
Attention-LSTM	98.43	97.63	94.55
BERT	98.69	98.15	94.36

5   Conclusions

In this chapter, three deep neural models have been used for the classification of Spanish news headlines about crimes from Twitter. Bidirectional LSTM, Attention-LSTM and BERT-based models start with free text extracted from criminal news and they use several layers to represent words (language) using weights, correlation between words and co-occurrences. Then, the output layer is used for determine the class label for a certain criminal news. We focus on six classes, such as: rape, kidnapping, homicide, suicide, assault and sexual exploitation. One of the objectives of this work is to determine the best model for our classification task.

The main contributions of this chapter are a) a proposed approach based on neural models to classify Spanish criminal news, b) the implementation and comparison three deep neural networks: Bidirectional LSTM, Attention-LSTM and BERT-based model, c) the experimentation and results of using two gold standard datasets from Spanish tweets and d) the conclusions about the best performance of BERT-based model for our classification task.

An evaluation process was carried out with two datasets as gold standard, extracted from Twitter and validated by humans. The three deep neural models are compared in terms of precision, recall and F_1. The results shows that the best neural network is the BERT-based model, which has achieved a 98.65% precision. Also, we can determine that the three neural models outperform the results shown as baseline.

In addition, the importance of work related to Natural Language Processing in Spanish must be highlighted since most of the works focus on English. Therefore, this chapter contributes to this gap and lack of NLP resources in the Spanish language. Furthermore, the task of classifying texts in Spanish will become more important in the future due to the increasing data available.

As future work, the word embeddings used can be changed to other models calculated through different methods or trained on another corpora, such as FastText [37] embeddings trained on Spanish untagged corpora or Glove [38] embeddings trained on Spanish Words Billion Corpus, to improve metrics results. Additionally, more tweets from different media news groups could be collected in order to test the proposed deep neural models. Finally, a system capable of classifying news according to some specific criteria such as crime news only could be useful for designing recommendation systems to digital newspaper readers or to extract knowledge (e.g. Named Entity Recognition or Relationship Extraction) for other tasks through text mining techniques.

Acknowledgments

This work is supported by the Sectoral Research Fund for Education with the CONACYT project 257357 and partially supported by the VIEP-BUAP project. The authors also would like to thank Universidad Autónoma Metropolitana, Azcapotzalco, with the research project SI001-18.

References

[1] C. C. Aggarwal and C. Zhai, A survey of text classification algorithms. in C. C. Aggarwal and C. Zhai (eds.), *Mining Text Data*. Springer, ISBN 978-

1-4419-8462-3, pp. 163–222 (2012), ISBN 978-1-4419-8462-3, `http://dblp.uni-trier.de/db/books/collections/Mining2012.html#AggarwalZ12b`.

[2] S. Hochreiter and J. Schmidhuber, Long short-term memory, *Neural Computation* **9**, 8, pp. 1735–1780 (1997), doi:10.1162/neco.1997.9.8.1735.

[3] F. A. Gers, N. Schraudolph and J. Schmidhuber, Learning precise timing with LSTM recurrent networks, *Journal of Machine Learning Research* **3**, pp. 115–143 (2002).

[4] R. Jing, A self-attention based LSTM network for text classification, *Journal of Physics: Conference Series* **1207**, p. 012008 (2019), doi:10.1088/1742-6596/1207/1/012008, `https://doi.org/10.1088%2F1742-6596%2F1207%2F1%2F012008`.

[5] J. Devlin, M.-W. Chang, K. Lee and K. Toutanova, BERT: Pre-training of deep bidirectional transformers for language understanding, in *Proceedings of the 2019 Conference of the North American Chapter of the Association for Computational Linguistics: Human Language Technologies, Volume 1 (Long and Short Papers)*. Association for Computational Linguistics, Minneapolis, Minnesota, pp. 4171–4186 (2019), doi:10.18653/v1/N19-1423, `https://www.aclweb.org/anthology/N19-1423`.

[6] U. Cervino, J. García, R. Calvo and H. Ceccatto, Automatic classification of news articles in Spanish, in *Actas del Congreso Argentino de Ciencias de Computación*. Argentina (2004).

[7] D. Y. Liliana, A. Hardianto and M. Ridok, Indonesian news classification using support vector machine, *World Academy of Science, Engineering and Technology, International Journal of Computer, Electrical, Automation, Control and Information Engineering* **5**, pp. 1015–1018 (2011).

[8] S. Bajaj and V. Goyal, Text News Classification System using Naïve Bayes Classifier, *Research Cell: An International Journal of Engineering Sciences* **3**, pp. 209–213 (2014), `http://ijoes.vidyapublications.com/paper/Vol13/39-Vol13.pdf`.

[9] G. Krishnalal, S. B. Rengarajan and K. G. Srinivasagan, A new text mining approach based on hmm-svm for web news classification, *International Journal of Computer Applications* **1**, pp. 103–109 (2010).

[10] I. Dilrukshi, K. de Zoysa and A. Caldera, Twitter news classification using svm, *8th International Conference on Computer Science & Education* , pp. 287–291 (2013).

[11] J. Nowak, A. Taspinar and R. Scherer, Lstm recurrent neural networks for short text and sentiment classification, in L. Rutkowski, M. Korytkowski, R. Scherer, R. Tadeusiewicz, L. A. Zadeh and J. M. Zurada (eds.), *Artificial Intelligence and Soft Computing*. Springer International Publishing, Cham, ISBN 978-3-319-59060-8, pp. 553–562 (2017), ISBN 978-3-319-59060-8.

[12] M. Ali Fauzi, A. Zainal Arifin, S. Christiano Gosaria and S. P. Isnan, Indonesian news classifica-tion using naïve bayes and two-phase feature selection model, *Indonesian Journal of Electrical Engineering and Computer Science* **8**, 3, pp. 610–615 (2017), doi:10.11591/ijeecs.v8.i3.pp610-615.

[13] C.-V. García-Mendoza and O. Gambino Juárez, News article classification of mexican newspapers, in M. F. Mata-Rivera and R. Zagal-Flores (eds.), *Telematics and Computing*. Springer International Publishing, Cham, ISBN 978-3-030-03763-5, pp. 101–109 (2018), ISBN 978-3-030-03763-5.

[14] R. Singh, S. A. Chun and V. Atluri, Developing machine learning models to automate news classification, in *The 21st Annual International Conference on Digital Government Research*, dg.o '20. Association for Computing Machinery, New York, NY, USA, ISBN 9781450387910, p. 354–355 (2020), ISBN 9781450387910, doi:10.1145/3396956.3397001, `https://doi.org/10.1145/3396956.3397001`.

[15] J. Liu, C. Xia, X. Li, H. Yan and T. Liu, A bert-based ensemble model for chinese news topic prediction, in *Proceedings of the 2020 2nd International Conference on Big Data Engineering*, BDE 2020. Association for Computing Machinery, New York, NY, USA, ISBN 9781450377225, p. 18–23 (2020), ISBN 9781450377225, doi:10.1145/3404512.3404524, `https://doi.org/10.1145/3404512.3404524`.

[16] R. Lin, C. Fu, C. Mao, J. Wei and J. Li, Academic news text classification model based on attention mechanism and rcnn, in Y. Sun, T. Lu, X. Xie, L. Gao and H. Fan (eds.), *Computer Supported Cooperative Work and Social Computing*. Springer Singapore, Singapore, ISBN 978-981-13-3044-5, pp. 507–516 (2019), ISBN 978-981-13-3044-5.

[17] C. Li, G. Zhan and Z. Li, News text classification based on improved bi-lstm-cnn, in *2018 9th International Conference on Information Technology in Medicine and Education (ITME)*, pp. 890–893 (2018).

[18] J. A. Reyes-Ortiz, Criminal event ontology population and enrichment using patterns recognition from text, *International Journal of Pattern Recognition and Artificial Intelligence* **33**, 11, p. 1940014 (2019), doi:10.1142/S0218001419400147.

[19] J. A. Reyes-Ortiz and M. Bravo, Enhancing patterns with linguistic information for criminal event recognition, *Journal of Intelligent and Fuzzy Systems* **34**, 5, pp. 3027–3036 (2018), doi:10.3233/JIFS-169487.

[20] A. Ritter, E. Wright, W. Casey and T. Mitchell, Weakly supervised extraction of computer security events from twitter, in *Proceedings of the 24th International Conference on World Wide Web*. Florence, Italy, pp. 896–905 (2015).

[21] B. Zhu, W. Yu and H. Cheng-Long, Global social event extraction and analysis by processing online news, in *Proceedings of the International Conference on Information System and Artificial Intelligence*. Hong Kong, pp. 73–76 (2016), doi:10.1109/ISAI.2016.0024.

[22] M. Atkinson, M. Du, J. Piskorski, H. Tanev, Y. R and V. Zavarella, Techniques for multilingual security-related event extraction from online news, in *Przepiórkowski A., Piasecki M., Jassem K., Fuglewicz P. (eds) Computational Linguistics*, Vol. 458. Sofia, Bulgaria, pp. 163–168 (2013), doi: 10.1007/978-3-642-34399-5_9.

[23] X. Yandi and L. Yang, Chinese event extraction using deep neural network with word embedding, in *arXiv preprint arXiv:1610.00842*. Sofia, Bulgaria (2016).

[24] X. Ding, Y. Zhang, T. Liu and J. Duan, Deep learning for event-driven stock prediction, in *Proceedings of the Twenty-fourth international joint conference on artificial intelligence*. Buenos Aires, Argentina, pp. 2327–2333 (2015).

[25] M. Xu, X. Zhang and L. Guo, Jointly detecting and extracting social events from twitter using gated bilstm-crf, *IEEE Access* **7**, pp. 148462–148471

(2019), doi:10.1109/ACCESS.2019.2947027.

[26] C. Cardellino, Spanish Billion Words Corpus and Embeddings, (2019), `https://crscardellino.github.io/SBWCE/`.

[27] S. Hochreiter, Y. Bengio, P. Frasconi and J. Schmidhuber, Gradient flow in recurrent nets: the difficulty of learning long-term dependencies, in S. C. Kremer and J. F. Kolen (eds.), *A Field Guide to Dynamical Recurrent Neural Networks*. IEEE Press (2001).

[28] A. Graves and J. Schmidhuber, Framewise phoneme classification with bidirectional lstm and other neural network architectures, *Neural Networks* **18**, 5-6, pp. 602–610 (2005).

[29] N. Srivastava, G. Hinton, A. Krizhevsky, I. Sutskever and R. Salakhutdinov, Dropout: A simple way to prevent neural networks from overfitting, *Journal of Machine Learning Research* **15**, 56, pp. 1929–1958 (2014), `http://jmlr.org/papers/v15/srivastava14a.html`.

[30] Y. Zhang and B. Wallace, A sensitivity analysis of (and practitioners' guide to) convolutional neural networks for sentence classification, in *Proceedings of the Eighth International Joint Conference on Natural Language Processing (Volume 1: Long Papers)*. Asian Federation of Natural Language Processing, Taipei, Taiwan, pp. 253–263 (2017), `https://www.aclweb.org/anthology/I17-1026`.

[31] A. Amidi and S. Amidi, Recurrent neural networks cheatsheet, (2020), `https://stanford.edu/~shervine/teaching/cs-230/cheatsheet-recurrent-neural-networks#attention`.

[32] A. Galassi, M. Lippi and P. Torroni, Attention in natural language processing, (2019), `arXiv:1902.02181 [cs.CL]`.

[33] J. A. Reyes-Ortiz and M. Bravo, Enhancing patterns with linguistic information for criminal event recognition, *Journal of Intelligent & Fuzzy Systems* **34**, 5, p. 3027 – 3036 (2018).

[34] J. Bergstra and Y. Bengio, Random search for hyper-parameter optimization. *J. Mach. Learn. Res.* **13**, pp. 281–305 (2012), `http://dblp.uni-trier.de/db/journals/jmlr/jmlr13.html#BergstraB12`.

[35] L. N. Smith, A disciplined approach to neural network hyper-parameters: Part 1 - learning rate, batch size, momentum, and weight decay, *CoRR* **abs/1803.09820** (2018), `arXiv:1803.09820`, `http://arxiv.org/abs/1803.09820`.

[36] A. S. Maiya, ktrain: A low-code library for augmented machine learning, (2020), `arXiv:2004.10703 [cs.LG]`.

[37] P. Bojanowski, E. Grave, A. Joulin and T. Mikolov, Enriching word vectors with subword information, *Transactions of the Association for Computational Linguistics* **5**, 0, pp. 135–146 (2017), `https://transacl.org/ojs/index.php/tacl/article/view/999`.

[38] J. Pennington, R. Socher and C. Manning, GloVe: Global vectors for word representation, in *Proceedings of the 2014 Conference on Empirical Methods in Natural Language Processing (EMNLP)*. Association for Computational Linguistics, Doha, Qatar, pp. 1532–1543 (2014), doi:10.3115/v1/D14-1162, `https://www.aclweb.org/anthology/D14-1162`.

https://doi.org/10.1142/9789811239014_0013

Chapter 13

Predicting US Elections with Social Media and Neural Networks

Ellison Yin Nang Chan, Adam Krzyżak[1] and Ching Y. Suen

Department of Computer Science and Software Engineering
Concordia University

[1]krzyzak@cs.concordia.ca

Abstract. Increasingly, politicians and political parties are engaging their electors using social media. In the US Federal Election of 2016, candidates from both parties made heavy use of Social Media, particularly Twitter. It is then reasonable to attempt to find a correlation between popularity on Twitter, and eventual popular vote in the election. In this study, we will focus on using the subscriber 'location' field in the profile of each candidate to estimate support in each state.

In this Study, we will train a Deep Convolutional Neural Network (CNN) to classify place names by state. Then we will apply the model to the Twitter Subscriber 'location' field of Twitter subscribers collected from each of the two candidates, Hillary Clinton (D), and Donald Trump (R). Finally, we will compare predicted popular votes in each state, to the actual results from the 2016 Presidential Election.

The hypothesis is that a city name has a strong correlation to the people who founded it and then incorporated it. Further, it is hypothesized that the original settlers were mostly homogeneous, relative to the country of origin and shared a common language, thus resulting in place names using the language of their origin. The results from our experiments are very promising.

Keywords: Election Forecasting, Deep Convolutional Neural Networks, Social Media

1   Introduction

1.1   Motivation and Research Topic

The goal is to use the locations of subscribers to each Candidate to determine who will win the most popular votes by state.

We will classify city and town names into states by using a suitably trained CNN classifier. It is felt from after having studied the analysis done by LeCun [5] that for its accuracy and shorter convergence time, especially when using GPU acceleration that CNN is an ideal algorithm for classifying the state names when representing them as sparse images.

Challenges

The Twitter "location" field in a Twitter subscriber profile provides information about where the user resides. However, the text is unconstrained. To obtain information from this Twitter field requires filtering out text that has a low likelihood of being a location, then grouping the remaining locations into US States. Since we have over 7.8 million locations to classify, we need an algorithm suitable for such a large dataset. We believe that a supervised algorithm, such as Convolutional Neural Network, will achieve the goal of training a model which can predict a state given a city name. Additionally, many modern Deep Learning frameworks such as TensorFlow, and Keras can use GPU acceleration to speed up training.

Methodology

In this study, our primary tool for learning the similarity function for state classification will be done using a Convolution Neural Network, with the Keras library running in the TensorFlow framework with GPU acceleration. After training a CNN to recognize an input dataset of cities labelled with their corresponding states, we will apply this CNN to predict the US state of a Twitter Subscriber, based on what they input into the 'location' field in their profile. We will then compare the predicted result to the actual election results for the 2016 US Federal Election. We will encode the cities as bigrams, with a vocabulary of 26 uppercase and 26 lowercase letters, including 3 non-alphabet characters, giving us a total vocabulary size of 55 characters. We will then convert this bigram matrix into 55x55 RGB images. This will enable a CNN to train on the city names. Use of CNN for NLP is not new. Typically, sentences are converted into vectors using word embedding, using an algorithm such as Word2Vec [2], from Google. Recently, Google has patented this algorithm, which might have an impact on the research community. In related work at CENPARMI, Ebrahimi, Mohammad Reza; Suen, Ching Y.; Ormandjieva, Olga [3], has used a Deep Learning CNN and Word2Vec with GPU acceleration, in order to detect predatory conversations, using social media. In the paper, "CNN-Webshell: Malicious Web Shell Detection with Convolutional Neural Network" [4], a CNN and Word2Vec are used to detect malicious Web Shell patterns inside the URLs of HTTP requests.

Novelty of Study Approach

In this study, we are interested in building a classifier for state names. We, therefore, need to apply context to the letters in relationship to specific state names. The feature we choose to use, is the frequency of occurrence of a bigram in city names, in the state. Each state will have a different frequency which could uniquely identify the states. An example of plots of bigrams frequency is seen in Fig. 1, which is the side view. The novelty of this Study approach is that rather than relying on Google's word embedding, we train a CNN to learn the bigram embedding in relationship to each state. We do this by training a CNN to recognize the affinity that a state has to a set of letter bigrams. It is believed that this a new approach that has not been tried yet.

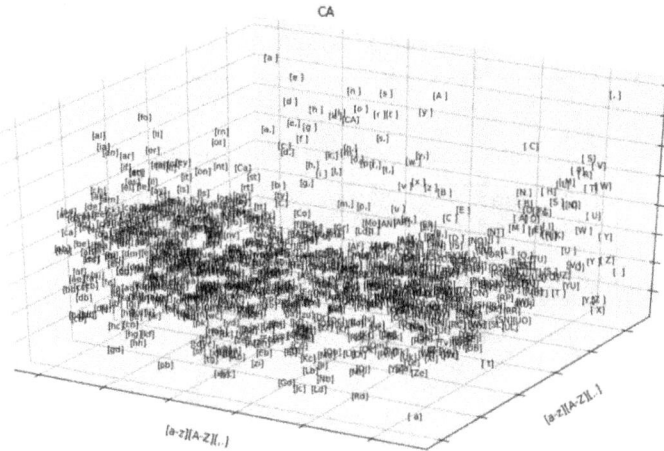

Fig. 1. California Bigram Frequency Plot (side view)

Data Description and Model Training

The method we decided to use to build our model with is a CNN for classifying our city names. With current advances in hardware (GPU training), and software (TensorFlow), we can train on GPU hardware and Deep Learning Neural Networks. We will be converting location names to bigram frequency, and them mapping them on two axes and using a Convolution Neural Network to classify into states by image recognition. In our algorithm, we choose to represent place names as bigrams and then convert bigrams into a two-dimensional matrix with each point representing the frequency of occurrence of a bigram in the place name.

Data Description

The data comes in two categories. The first is a labelled set of city names and labelled with each state that the city belongs to. These are all correctly spelled and correctly labelled. This is the basis of the training, validation, and test sets, from which the model is trained, validated, and tested. A sample of this is found in Table 1. We will place it under the umbrella of Model Data.

Table 1. US Cities Dataset (Excerpt)

```
1,Adak,Aleutians West Census Area,AK,Alaska,City,51.88,-176.65806
2,Akhiok,Kodiak Island Borough,AK,Alaska,City,56.94556,-154.17028
3,Akiachak,Bethel Census Area,AK,Alaska,CDP,60.90944,-161.43139
4,Akiak,Bethel Census Area,AK,Alaska,City,60.91222,-161.21389
5,Akutan,Aleutians East Borough,AK,Alaska,City,54.13556,-165.77306
6,Alakanuk,Kusilvak Census Area,AK,Alaska,City,62.68889,-164.61528
7,Alatna,Yukon-Koyukuk Census Area,AK,Alaska,CDP,66.56393,-152.838
```

Table 2. Sample Twitter Profile Locations

```
[{"logged_at": "2016-10-23T10:30:06",
 "created_at": "2011-12-09T18:30:19",
 "id_str": "433007427",
 "last_tweet": "788554771250581505",
 "favourites_count": 324,
 "followers count": 56,
 "friends count": 184,
 "listed count": 3,
 "statuses count": 294,
 "following_id_str": "HillaryClinton",
 "is_translator": 0,
 "geo_enabled": 0,
 "location": "Arlington, Texas",
 "verified": 0,
 "screen_name": "<removed>",
 "lang": "es",
 "utc_offset": -1,
 "time_zone": "", "name":
 "????", "description":
 "<removed>",
 "argmax": 0.0,
 "orig": null,
 "idx": null,
 "prob": null,
 "correct": null}]
```

The second category of input data is the Twitter subscriber profile 'location' field. This is the data which we want to apply the model against to help us classify Twitter subscribers into states for each of the election candidates. A sample of these data can be found in Table 2. The data has been converted to JSON format with only the essential information that we anticipate will be needed for training and prediction. (Table 3) We will extract the location of each subscriber and convert the location field into bigrams. (Table 4).

Table 3. US Cities Dataset Converted o JSON

```
{"state_short": "FL", "letters": ["FL", " W", "at", "er",
"ga", "te"], "location": "FL Watergate"}
{"state_short": "TX", "letters": ["Cr", "ec", "y,", " T",
"X "], "location": "Crecy, TX"}
{"state_short": "TN", "letters": ["Da", "nd", "ri", "dg",
"e ", "TN"], "location": "Dandridge TN"}
{"state_short": "NC", "letters": ["NC", " L", "um", "be",
"rt", "on"], "location": "NC Lumberton"}
```

Table 4. Sample Twitter Profile Locations Converted to JSON

```
{"letters": ["so", "me", "wh", "er", "e ", "be", "in", "g
", "pe", "tt", "y "], "following_id_str": "HillaryClinton",
"logged_at": "2016-10-23T10:30:06", "id_str": "35399372",
"created_at": "2009-04-25T22:11:25", "verified": 0}
{"letters": ["Ar", "li", "ng", "to", "n,", " T", "ex",
"as"], "following_id_str": "HillaryClinton", "logged_at":
"2016-10-23T10:30:06",      "id_str":      "859131001",
"created_at": "2012-10-02T21:35:56", "verified": 0}
{"letters": ["DC"], "following_id_str": "HillaryClinton",
"logged_at":      "2016-10-23T10:30:06",      "id_str":
"2333943649",    "created_at":    "2014-02-08T12:02:58",
"verified": 0}
```

Model Input

We apply a permutation of place names to augment the input data, and remove any duplicate city names, the total number of cities with corresponding state names is 695, 389 entries. We will calculate a letter bigram for each city, and then count the frequency of occurrence in each city name.

Algorithm

The bigrams for each place name will be converted into an image representation to be fed into a CNN for training.

1.1.1 Vocabulary of Characters

We want to limit our vocabulary size for our input data so that each city name will not map into too large a matrix for the efficient training and running of our model. Characters chosen to be part of our vocabulary are listed in Table 5. We will include two punctuations, comma and period, which are important in place names. Space is included in the vocabulary, but we will trim the left and right of the place name and remove punctuations and spaces. Capital letters being important to places names are retained. Only internal punctuation and spaces are kept. With the x-axis representing bigram combinations of our vocabulary and. y-axis representing bigram frequency shows histogram plots of bigram frequencies for some sample states. The question this study tries to answer is whether a neural network can be generalized to the recognition of city names, which are not a part of the training data.

Table 5. Characters in the Vocabulary

Lower Case Letters	abcdefghijklmnopqrstuvwxyz
Upper Case Letters	ABCDEFGHIJKLMNOPQRSTUVWXYZ
Punctuation (internal only)	,.
Space (internal only)	

1.1.2 Histogram Representation

Our bigrams can be represented by a 55x55 matrix of bigram frequencies, giving us a mapping space of 3025 bigrams forming a 3D histogram. Histogram algorithms have been successfully applied in previous work to recognize sparse featured images such as handwriting. In this paper the authors used it for Arabic handwriting. [8]

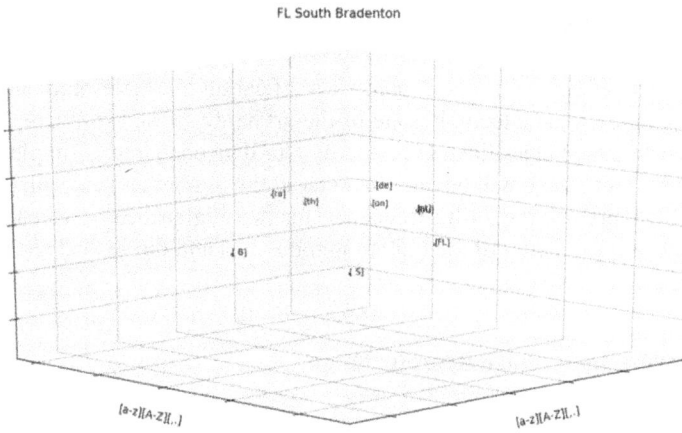

Fig. 2. Bigram Frequency of South Bradenton, FL(Side)

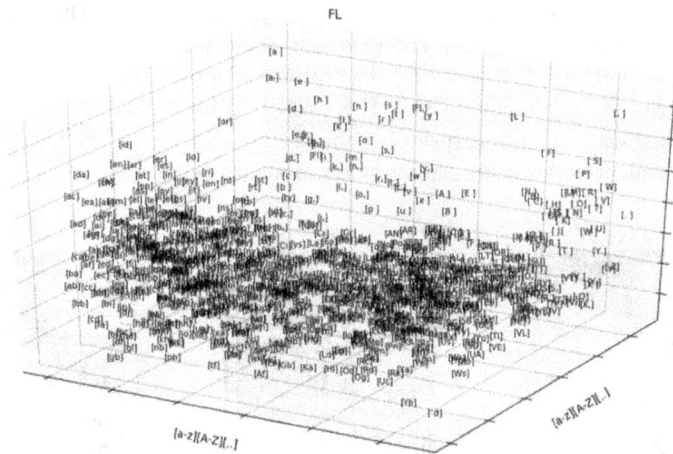

Fig. 3. Bigram Frequency for Florida (Side)

1.1.3 Data Encoding

We want to reframe the place name classification problem into an image processing problem, and to apply state-of-the-art hardware and software, actively being used in image classification research. Our data encoding is simple. The x and y-axis of our image will be one character in the bigram and each pixel in our images will represent the 24-bit frequency for that bigram. By convention, we will encode the first letter of the bigram in the x-axis and the second character in the y-axis.

1.1.4 Converting to Image Representation

To enable a CNN to train on our histograms, we will convert them into a two-dimensional image. The x and y-axis are indexes to each letter of our vocabulary, and the RGB values represent a 24-bit sized frequency of occurrence for the bigram. Sample images built in this manner are shown in Fig. 4. The images help to illustrate that some bigrams combinations occur more frequently with some states. Together this represents a "fingerprint" of sorts that we hope a neural network can learn the pattern to. The CNN will be trained to recognize the features that each state fingerprint is composed of. In Fig. 4, are the images of bigram frequencies for some sample states individually.

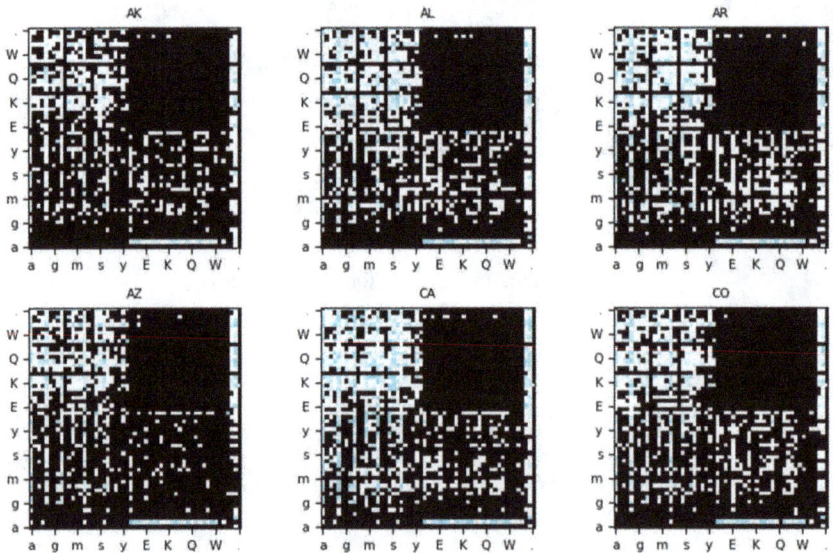

Fig. 4. Image Representation of Bigram Frequencies (Sample)

1.1.5 Augmenting the Dataset

To augment the basic list of city names and states, the operations in Table 6 are applied to the original list. This will help to capture the variations that people will use in the location field of their Twitter profile. It is interesting to note that these permutations do not contain the city name alone with no state name. This is done intentionally to avoid confusing the neural network. Many states carry common city names and adding this permutation would result in adjusting weights towards one state, and then when the same city came in for another state it would undo that training resulting in the neural network oscillating back and forth, until the last entry of that city.

Table 6. Permutation Operation on State Names and Cities

Operation	Description
[Long State Name]	Long state name alone
[City], [Long State Name]	City and Long State separated by a comma
[City] [Long State Name]	City and Long State Name separated by space
[Long State Name], [City]	Long State Name and City separated by a comma
[Long State Name] [City]	Long State Name and City separated by a space
[City], [Short State Name]	City and Short State Name separated by a comma
[City] [Short State Nam]	City and Short State name separated by a space
[Short State Name], [City]	Short State Name and City separated by a comma
[Short State Name] [City]	Short State Name and City separated by a space

1.1.6 Shuffling Data

An initial shuffling of the data is necessary, for the next step of splitting between the training, validation, and test sets. This will ensure an even distribution for each of the datasets after splitting.

1.1.7 Splitting Data into Training, Validation, and Test Sets

The final step is to split the data into the separate components. We will keep 80% of the original data for training while splitting the remaining 20% into validation (16%), and test (4%).

Table 7. Training, Validation, Test Data

	City Entries	File Name	Description Statistics Numpy.describe()
Full Data (100%)	695389	cities_shuffled_all.json	nobs=52, minmax=(0, 46271), mean=12929.98076923077, variance=94827083.3525641, skewness=1.3722954619398018, kurtosis=1.9301177227759876
Training (80%)	556311	cities_shuffled_train.json	nobs=52, minmax=(0, 37041), mean=10342.307692307691, variance=60726949.62895928, skewness=1.3738867581234702, kurtosis=1.9387584597763698
Validation (16%)	111262	cities_shuffled_valid.json	(nobs=52, minmax=(0, 7420), mean=2069.6923076923076, variance=2435641.3152337857, skewness=1.3721646821298885, kurtosis=1.915920922857353
Test (4%)	27816	cities_shuffled_test.json	nobs=52, minmax=(0, 1810), mean=517.9807692307693, variance=148579.43099547512, skewness=1.3356616905991185, kurtosis=1.7888861304572803

Model Preparation

1.1.8 Model Definition

We are using five 2D convolution layers with 2x2 max pooling layers between each convolution layer. The Keras Neural Network Framework is being used to define our CNN. Keras is a popular framework for defining, training, and generating predictions, and provides GPU acceleration using a TensorFlow Backend. The programming language of choice is Python. The model is summarized in Table 8. We will provide the details in the next sections.

Table 8. Keras Model Definition

Layer (type)	Output Shape	Param #
conv2d_1 (Conv2D)	(None, 55, 55, 32)	896
_activation_1 (Activation)	(None, 55, 55, 32)	0
_batch_normalization_1	(Batch (None, 55, 55, 32)	128
_max_pooling2d_1 (MaxPooling2)	(None, 18, 18, 32)	0
_dropout_1 (Dropout)	(None, 18, 18, 32)	0
_conv2d_2 (Conv2D)	(None, 18, 18, 64)	18496
_activation_2 (Activation)	(None, 18, 18, 64)	0
_batch_normalization_2	(Batch (None, 18, 18, 64)	256
_conv2d_3 (Conv2D)	(None, 18, 18, 64)	36928
_activation_3 (Activation)	(None, 18, 18, 64)	0
_batch_normalization_3	(Batch (None, 18, 18, 64)	256
_max_pooling2d_2 (MaxPooling2)	(None, 9, 9, 64)	0
_dropout_2 (Dropout)	(None, 9, 9, 64)	0
_conv2d_4 (Conv2D)	(None, 9, 9, 128)	73856
_activation_4 (Activation)	(None, 9, 9, 128)	0
_batch_normalization_4 (Batch)	(None, 9, 9, 128)	512
_conv2d_5 (Conv2D)	(None, 9, 9, 128)	147584
_activation_5 (Activation)	(None, 9, 9, 128)	0
_batch_normalization_5 (Batch)	(None, 9, 9, 128)	512
_max_pooling2d_3 (MaxPooling2)	(None, 4, 4, 128)	0
_dropout_3 (Dropout)	(None, 4, 4, 128)	0
_flatten_1 (Flatten)	(None, 2048)	0
_dense_1 (Dense)	(None, 1024)	2098176
_activation_6 (Activation)	(None, 1024)	0
_batch_normalization_6 (Batch)	(None, 1024)	4096
_dropout_4 (Dropout)	(None, 1024)	0
_dense_2 (Dense)	(None, 52)	53300
_activation_7 (Activation)	(None, 52)	0

Model Training

We trained our CNN model on 556,311 city names from our training dataset. In addition to the model definition in Table 8, when training there are hyperparameters that need to be set, which can affect the outcome of the trained model. These are listed in Table 9. We will discuss these parameters in greater detail and provide some discussion on how training and model accuracies are affected in the next sections.

Table 9. Training Hyperparameters

Parameter	Value	Description
Batch Size	32	Number of samples between gradient calculation.
Optimizer	Adam	Adam optimizer, with an initial learning rate of 0.001, and decay rate of 0.001.
Loss Function	Binary Cross Entropy	Calculates the loss between each gradient calculation.
Maximum Epochs	<= 75	Rather than using a fixed number of epochs, we are using early stop, if the validation loss does not change for 4 epochs.

1.1.9 GPU Acceleration

The strength of GPU acceleration is that we can calculate the optimization and losses faster and use a smaller batch size. Each epoch in the training takes approximately 40 minutes per epoch, running on a Nvidia GTX-1070 video card. Without GPU acceleration, training with such a small block size would take approximately 75 minutes per epoch running on an Intel I7 processor with 16GB of system memory. This represents an almost 2:1 ratio. Our CNN model took 12 epochs to converge. This translates to 8 hours to train the model on the full training set on GPU versus 15 hours on a CPU. That is a significant reduction in training time.

Model Cross-Validation

We will use 10-Fold Cross-Validation to evaluate our CNN classifier's performance. The full training dataset is split into 10-folds, with 1-fold held out as training data, while the model is trained on the other 9 folds.

1.1.10 Randomized Data

To ensure we each fold is a good representation of the total dataset, to avoid any skewing during validation, we have compared the histograms of city and state frequencies in the full dataset.

1.1.11 Cross-Validation Results

The mean Cross-Validation accuracy is 99.74% with a standard deviation of +/-0.00. This is an indication that the model behaves stably. As with the full training dataset, we are only limiting cross-validation to the 51 states which are eligible to vote in the 2016 Presidential Election.

Discussion

We will look at the Model applied to each of the Datasets (Training, Validation, and Test) to gauge the model's accuracy before applying it to classify Twitter subscriber "location" field, to predict the election outcome. To judge the accuracy of the model on new data, we will perform some standard tests, including a T-Test to gauge the variance between predicted election results versus actual results. Finally, we will look at each state's election results and determine our accuracy in predicting election results for the two candidates.

Analytical Approach

First, we will run the model on the Test Dataset which we partitioned off from the main data. We will do a quantitative analysis by calculating the prediction rates and plotting for each dataset by state. Prediction accuracy for the test data set is summarized in Table 10.

Table 10. Prediction Results

Prediction Type	Test	Training	Validation	Description
False Positives	1.0700%	1.583%	1.708%	Predictions with a probability greater than and equal to 50%, where the labelled state does not match with the predicted state.
False Negatives	1.611%	2.007%	11.551%	Predictions with a probability less than 50%, where the labelled state matches the predicted state.
Wrong Prediction	12.263%	11.495%	12.074%	Predictions with probability less than 50%, where the labelled state and predicted state do not match. (Note that this means the model assigned a state but was able to determine that there was a low probability.)
Correct Prediction	84.437%	84.914%	84.667%	Predictions with probability greater than and equal to 50%, where labelled and predicted states match.

Predicting 2016 US Election with Trained CNN Model

Now that we have trained a CNN model to classify locations into US states, we can apply the model to predict the election results for the Republican and Democratic candidates in the 2016 US Federal Election.

Results Analysis

1.1.12 T-Test of Predicted vs Actual Results

Comparing both candidates Trump and Clinton predictions to their respective actual results show that the variance between prediction and actual are not equal. Because of this, we will apply the T-Test for unequal variances, using Microsoft Excel, for the two samples. In candidate Trump's case, the t-stat of 0.46517 is between the range of +/- t-critical value 1.99444. According to this result, we can say that the average difference between predicted results versus actual results is not dissimilar. Within statistical significance, the model has predicted a match with the actual result. Similar results are had for Clinton. The t-stat value -0.46517 falls between t-critical range of -1.99444 and 1.99444. Both F-tests and T-tests have similar results indicating that the model is stable between the two Candidate datasets

1.1.13 Election Results Predicted vs Actual

For our Election Results analysis, we drew from two sources. The first one, Source #1, contains popular votes for all candidates and parties by state and county.   (https://data.opendatasoft.com/explore/dataset/usa-2016-presidential-election-by-county@public/).

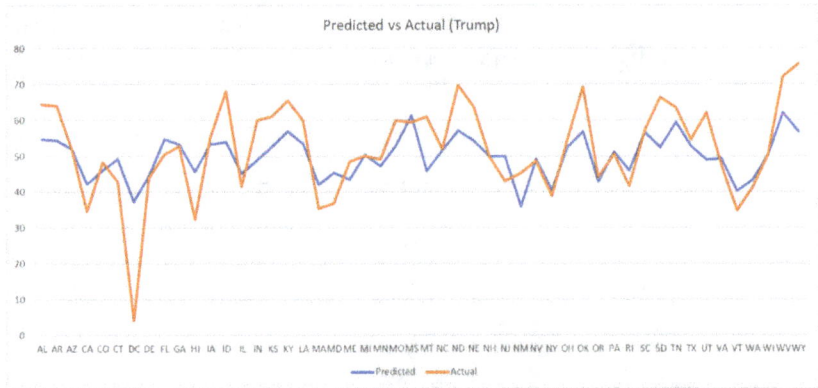

Fig. 5. Predict vs Actual (Trump)

We will use this source as the basis of our popular vote comparison by state. Of note is that this source was missing vote counts for the state of Alaska, so the analysis did not include this state. The second one, Source #2, give us a second sanity check and provides the total popular vote by each candidate. https://www.270towin.com/ 2016_Election/

We will resample the Twitter subscriber data so that the total Trump subscribers versus total Clinton subscribers will match closely with this ratio.

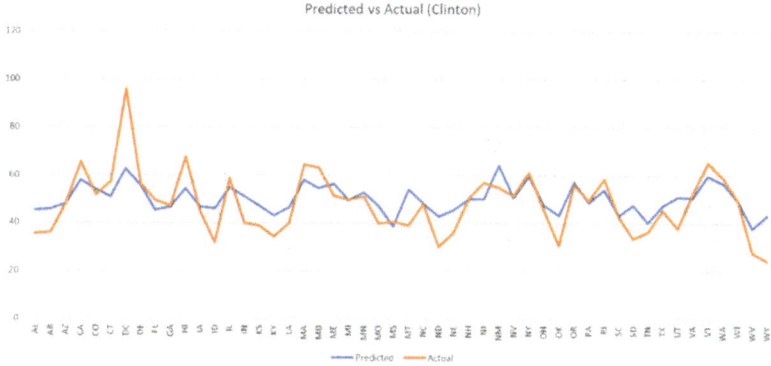

Fig. 6. Predict vs Actual (Clinton)

Finally, the model predicts overall that Trump won 29 states, while Clinton was predicted to win 21.

	Predict Sample	Actual (Source1)	Actual (Source2)
Trump	387403	61064602	62980160
Clinton	404738	62426228	65845063
Percentage	48.91%	49.45%	49.45%

Fig. 7. Sample Ratio

Column1	# States (Pred)	# States (Act)
Trump Wins	28	29
Clinton Wins	22	21

Fig. 8. Overall Winner

Matched	45
Mismatched	5
% Matched	90%

Fig. 9. Prediction Rate for Election 2016

2 Conclusion

We feel this project has been successful. The major finding is that we can reframe a traditional text pattern matching task into an image recognition task. We have also successfully applied a trained machine learning model to help predict locations of Twitter subscribers of the two major candidates in the 2016 US general election, and have our predicted results match actual results. However, the model needs to be refined further to prevent breaking down, in cases that location names do not carry enough clues for classification.

References

1. C. Cortes, C. J. Burges and Y. LeCun, "MNIST database of handwritten digits - Yann LeCun," [Online]. Available: http://yann.lecun.com/exdb/mnist/.
2. Y. Li, K. Zhao, X. Chu and J. Liu, "Speeding up k-Means algorithm by GPUs," Journal of Computer and System Sciences, vol. 79, no. 2, pp. 216–229, 2013.
3. Q. V. Le and T. Mikolov, "Distributed Representations of Sentences and Documents," in Proceedings of the 31st International Conference on International Conference on Machine Learning - Volume 32, Beijing, China, pp. 1188–1196, 2014.
4. M. R. Ebrahimi, C. Y. Suen and O. Ormandjieva, "Detecting predatory conversations in social media by deep Convolutional Neural Networks," Digital Investigation, vol. 18, pp. 33–49, 2016.
5. Y. Tian, J. Wang, Z. Zhou and S. Zhou, "CNN-Webshell: Malicious Web Shell Detection with Convolutional Neural Network," in Proceedings of the 2017 VI International Conference on Network, Communication and Computing, Kunming, China, pp. 75–79, 2017.
6. J. Fei, H. Fang, C. Yang and D. Wang, "Restricted Stochastic Pooling for Convolutional Neural Network," in Proceedings of the 10th International Conference on Internet Multimedia Computing and Services, Nanjing, China, pp. 1–4, 2018.
7. S. Park and N. Kwak, "Analysis on the Dropout Effect in Convolutional Neural Networks," in Computer Vision—ACCV 2016, Taipei, Taiwan, pp. 189–204, 2016.
8. N. A. Jebril, H. R. Al-Zoubi and Q. A. Al-Haija, "Recognition of Handwritten Arabic Characters using Histograms of Oriented Gradient (HOG)," Pattern Recognition and Image Analysis, vol. 28, no. 2, pp. 321–345, 2018.

9. S. Ioffe and C. Szegedy, "Batch Normalization: Accelerating Deep Network Training by Reducing Internal Covariate Shift," in ICML'15 Proceedings of the 32nd International Conference on International Conference on Machine Learning, Lille, France, pp. 448–456, 2015.

10. Y. Wang, X. Wu, Y. Chang, S. Zhang, Q. Zhou and J. Yan, "Batch Normalization: Is Learning An Adaptive Gain and Bias Necessary?," in Proceedings of the 2018 10th International Conference on Machine Learning and Computing, Macau, China, pp. 36–40, 2018.

11. K. Gopalakrishnan, S. K. Khaitan, A. N. Choudhary and A. Agrawal, "Deep Convolutional Neural Networks with transfer learning for computer vision-based data-driven pavement distress detection," Construction and Building Materials, vol. 157, pp. 322–330, 2017.

12. F. Xiong, Y. Xiao, Z. Cao, K. Gong, Z. Fang and J. T. Zhou, "Towards Good Practices on Building Effective CNN Baseline Model for Person Re-identification.," Computer Vision and Pattern Recognition, vol. abs/1807.11042, 2018.

13. R. Ha, P. Chang, J. Karcich, S. Mutasa, R. Fardanesh, R. Wynn, M. Z. Liu and S. Jambawalikar, "Axillary Lymph Node Evaluation Utilizing Convolutional Neural Networks Using MRI Dataset," Journal of Digital Imaging, vol. 31, no. 6, pp. 851–856, 2018.

14. H. Li, X. Gong, H. Yu and C. Zhou, "Deep Neural Network Based Predictions of Protein Interactions Using Primary Sequences," Molecules, vol. 23, no. 8, p. 1923, 2018.

15. S. Mannor, D. Peleg and R. Y. Rubinstein, "The cross entropy method for classification," in Proceedings of the 22nd International Conference on Machine Learning, Bonn, Germany, pp. 561–568, 2005.

16. S. Hallmark, R. R. Souleyrette and S. Lamptey, "Use of n-Fold Cross-Validation to Evaluate Three Methods to Calculate Heavy Truck Annual Average Daily Traffic and Vehicle Miles Traveled," Journal of The Air & Waste Management Association, vol. 57, no. 1, pp. 4–13, 2007.

17. P. Klęsk, "Probabilities of discrepancy between minima of cross-validation, Vapnik bounds and true risks," International Journal of Applied Mathematics and Computer Science, vol. 20, no. 3, pp. 525–544, 2010.

18. T. He, W. Huang, Y. Qiao and J. Yao, "Text-Attentional Convolutional Neural Network for Scene Text Detection," IEEE Transactions on Image Processing, vol. 25, no. 6, pp. 2529–2541, 2016.

19. Y. LeCun, "Gradient-based learning applied to document recognition," Proceedings of the IEEE, vol. 86, no. 11, p. 2278–2324, 1998.

20. B. Wang, L. Najjar, N. N. Xiong and R. C. Chen, "Stochastic Optimization: Theory and Applications," Journal of Applied Mathematics, vol. 2013, pp. 1–2, 2013.

21. J. Ding, X.-H. Hu and V. N. Gudivada, "A Machine Learning Based Framework for Verification and Validation of Massive Scale Image Data," IEEE Transactions on Big Data, vol. 7, no. 2, pp. 451–467, 2021.

22. N. Gajhede, O. Beck and H. Purwins, "Convolutional Neural Networks with Batch Normalization for Classifying Hi-hat, Snare, and Bass Percussion Sound Samples," in Proceedings of the Audio Mostly 2016, New York, NY, USA, pp. 111–115, 2016.

Chapter 14

Differences and Similarities learning for Unsupervised Feature Selection

Tan Jun[1], Xinyi Li[2], Ning Bi[3] and You Ning[4]

School of Mathematics, Sun Yat-Sen University
Guangzhou, 510275, P. R. China
Guangdong Province Key Laboratory of Computational Science
Sun Yat-Sen University, Guangzhou, 510275, P. R. China
[1]mcstj@mail.sysu.edu.cn, [2]390860549@qq.com
[3]mcsbn@mail.sysu.edu.cn, [4]youn7@mail2.sysu.edu.cn

Abstract: In this chapter, a novel feature selection algorithm, named Feature Selection with Differences and Similarities (FSDS), is proposed. FSDS jointly exploits sample differences from global structure and similarities from local structure. To reduce the disturbance from noisy feature, a row-wise sparse constraint is also merged into the objective function. FSDS, then combines the underlying subspace features with original feature to construct a more reliable feature set. Furthermore, a joint version of FSDS (FSDS2) is introduced. To optimize the proposed two-step FSDS and the joint version FSDS2 we also design two efficient iterative algorithms. Experimental results on various datasets demonstrate the effectiveness of the proposed algorithms.

Keywords: Feature Selection, Spectral Learning, Row Sparsity, Sample Differences and Similarities Exploiting

1   Introduction

The dilemma of high-dimension data i.e. time and space pressure, has brought about great challenges to various areas, including computer vision, pattern recognition and biological science. Moreover, the information redundancy together with noise in original data makes it significantly necessary to implement feature selection and dimension reduction.

Recent years have witnessed the increase on creative feature selection algorithms proposed to address dimension dilemma. Generally, they can be classified into three groups, 1) supervised algorithm, 2) semi-supervised algorithms and 3) unsupervised algorithms. Supervised feature selection methods

[1], [2] select important features by considering the correlation between features and label information. Supervised methods are comparatively efficient since the goal information is remaining in labels. However, labeling data requires excessive time and human labor. Semi-supervised feature selection methods discuss about both labeled data and unlabeled data based on some regularization frameworks, such as normalized min-cut with NMI classification loss [3] and semi-supervised SVM with manifold regularization [4]. However, atypical labeled data may fail to represent data structure and yield negative effect to feature selection procedure. In lack of sufficient label information, it is promising and sensible to develop unsupervised feature selection methods.

In unsupervised scenarios, feature selection is achieved according to different criteria (e.g. [5-6], [8-10], [12-14]), such as maximum variance. However, it ignores locality preserving whereas the local structure of the data space is more important than the global structure as mentioned in [6]. For this reason, many approaches utilizes a transformation matrix which can map the original space into a lower-dimensional one and simultaneously model the local geometric structure of the original space. Among these methods, some [6], [8-10] emphasize similarity preserving, which is achieved by preserving manifold structure of the original data space. Others [5] focuses on exploiting features that involves the most discriminative information between samples. Nevertheless, taking either sample similarities or differences into account may not be comprehensive enough and thus this two procedures will be joined in the proposed method. Additionally, it is conducive that feature selection should be well-directed and more label information should be assimilated into the procedure of feature selection, revealing the goal of final classification or clustering. For example, reference [5], [10], [12-14] employ pseudo labels to select effective features. To strengthen the effect of label information, not only pseudo labels are applied in this chapter, but k-means algorithm will also be utilized since the parameter k is related to experiment target to some extent. What's more, most references [8-10], [12-14] unify subspace learning technique into feature selection. There is no reason to assume that separating those two procedures must gain worse experimental performance. In this chapter, interesting experiments are conducted to explore the two situations.

The main contribution of this study relies on four aspects:

- We propose a new algorithm named Feature Selection with Differences and Similarities (FSDS) by simultaneously exploiting the sample differences from the global structure and the sample similarities from the local structure.
- Our method (FSDS) also considers selected original space feature and subspace features independently.
- We introduce a unifying framework of a "two-step" FSDS, named as FSDS2, in which the information of subspace features and original features fuses together, compared to FSDS.

- Corresponding iterative optimization algorithms of FSDS and FSDS2 are also proposed respectively in our work.

The remaining of this chapter is arranged as follows. We elaborate our proposed formulation FSDS in Section 2. Then optimization algorithm is presented in Section 3 followed with discussions about algorithm FSDS2 introduced in Section 4. Extensive experiments are conducted and analyzed in Section 5. Section 6 illustrates the conclusion of our work.

1.1   Preliminary

To apply a style, begin by selecting the text to which the style needs to be applied, and then choose the style name from the Styles gallery. The WSPC template contains all the styles that are required for formatting the documents. It is crucial and recommended to exploit the usage of styles as much as possible to format the text. However, apply direct formatting only as a last resort when the style, for some reason, is incompatible with the requisite. As a note of caution, do not use any style for a purpose other than that for which it was intended or for the want of effort.

In this section, we first summarize some notations used throughout this chapter. We use bold uppercase characters to denote matrices, bold lowercase characters to denote vectors. For an arbitrary matrix A, a_i means the i-th row vector of A, a^j means the j-th column vector of A, A_{ij} denotes the (i, j)-th entry of A and tr[A] is the trace of A if A is square. The $l_{r,k}$ norm of matrix A is defined as

$$\|A\|_{r,k} = \left(\sum_{i=1}^{r}\left(\sum_{j=1}^{t} A_{i,j}^{r}\right)^{k/r}\right)^{1/k} \tag{1.1}$$

Moreover, $\|A\|_F$ is Frobenius norm of A, formulated as $\|A\|_F = \left(\sum_{i=1}^{n}\sum_{k=1}^{m} A_{ik}^{2}\right)^{1/2}$. And l_r norm of vector v is defined as $\|v\|_r = \left(\sum_{i=1}^{n}|v_i|^r\right)^{1/r}$.

Assume that we have n samples $\chi = \{x_i\}_{i=1}^{n}$. Let $X = [x_1, \ldots, x_n] \in \mathcal{R}^{d \times n}$ denote the data matrix, in which $x_i \in \mathcal{R}^{d \times n}$ represents the i-th sample. Suppose these n samples are sampled from c classes and there are n_i samples in the i-th class. In this chapter, I is identity matrix, $1_m \in \mathcal{R}^m$ is a column vector with all elements being 1 and H_n is used for data centralizations, as in equation 3), $\tilde{X}$ is the data matrix after being centered.

$$H_n = I - \frac{1}{n}1_n 1_n^{T} \in \mathfrak{R}^{n \times n} \tag{1.2}$$

$$\tilde{X} = XH_n \tag{1.3}$$

We define $Y = [y_1, \ldots, y_n]^T \in \{0,1\}^{n \times c}$ to be the label matrix, which means that the j-th element of y_i is 1 if sample x_i belongs to the j-th class, and 0 otherwise. The scaled label matrix G is defined as

$$G = [G_1, \ldots, G_n]^T = Y(Y^T Y)^{-1/2} \in \Re^{n \times c} \tag{1.4}$$

It turns out that

$$G^T G = (Y^T Y)^{-1/2} Y^T Y (Y^T Y)^{-1/2} = I_c \tag{1.5}$$

The intraclass scatter matrix S_e and between class scatter matrix S_b are defined as follows.

$$S_e = \sum_{i=1}^{c} \sum_{x_j \in class\ i} (x_j - \mu_i)(x_j - \mu_i)^T = \tilde{X}(I_n - GG^T)\tilde{X}^T \tag{1.6}$$

$$S_b = \sum_{i=1}^{c} n_i (\mu_i - \mu)(\mu_i - \mu)^T = \tilde{X}GG^T \tilde{X}^T \tag{1.7}$$

where μ is the mean of all samples, μ_i is the mean of samples from the j-th class.

1.2 Related Works

Over the past several decades, many methods for dimension reduction have been raised to distinguish which feature set can best maintain the intrinsic clusters information in data set. Then, we will elaborate some dimensionality reduction methods connected with this chapter, such as Maximum Margin Criterion (MMC) [23], Linear Discrimination Analysis (LDA) [26], Multi-Cluster Feature Selection (MCFS) [8], Nonnegative Discriminative Feature Selection (NDFS) [10] and Unsupervised Discriminative Feature Selection (UDFS) [5].

Linear Discrimination Analysis (LDA) is a classical feature extraction method whose main idea is to learn the projection matrix W and get the label indicator matrix $F = X^T W$, guaranteeing that samples within each classes are close to each other whereas samples from different classes are far away from each other[26]. The objective function is formulated as:

$$\max\ J(w) = \frac{w^T S_b w}{w^T S_e w} \Leftrightarrow \max\ J(W) = \frac{tr[W^T S_b W]}{tr[W^T S_e W]} \tag{1.8}$$

where S_e is the within intraclass scatter matrix and S_b is the between class scatter matrix. To solve the problem (1.8), there are two kinds of methods: ratio trace and trace ratio [24]. Ratio trace transforms problem (1.8) into the following formula:

$$W^* = \arg\max_{W}\ tr[(W^T S_e W)^{-1}(W^T S_b W)] \qquad (1.9)$$

Moreover, (1.9) is equal to the equation below:

$$\min_{w}\ -w^T S_b w \ \ s.t.\ \ w^T S_e w = 1 \qquad (1.10)$$

Through solving generalized characteristic equation $S_b w_k = \beta_k S_e w_k$ and gaining eigenvectors w_k with the largest eigenvalues, the projection matrix W can be obtained, $W = [w^1, \ldots, w^c] \in R^{d \times c}$. Ratio trace problem is based on greedy algorithm [24] which iteratively searches for the eigenvector w_k that maximizes $(w^i)^T S_b w^i / (w^i)^T S_e w^i$ and   thus   essentially   optimizes $\sum_{i=1}^{c}(w^i)^T S_b w^i / (w^i)^T S_e w^i$. In contrast, the trace ratio problem [24] aims to optimize $\sum_{i=1}^{c}(w^i)^T S_b w^i / \sum_{i=1}^{c}(w^i)^T S_e w^i$. Therefore, the trace ratio problem can be written as:

$$W = \arg\max_{W^T W = I} \frac{tr[W^T S_b W]}{tr[W^T S_e W]} \qquad (1.11)$$

However, the problem (1.11) does not have a closed-form solution and needs to be transformed into trace difference problem [24]. According to Ref. [25], when $J(W)_{W^T W = I} = tr[W^T S_b W] / tr[W^T S_e W]$ reaches the maximum value λ_0, the equation $\max_{W^T W} tr[W^T(S_b - \lambda_0 S_e)W]$ will reach 0, and vice versa. In this way, the problem (1.11) is converted into trace difference problem, that is, to calculate the zero point of $f(\lambda) = \max_{W^T W} tr[W^T(S_b - \lambda S_e)W]$.

Besides, **Maximum Margin Criterion (MMC)** is another feature extraction algorithm, which is the special case of trace ratio problem[23]. MMC problem can be formulated as:

$$W = \arg\max_{W^T W = I}\ tr[W^T(S_b - S_e)W] \qquad (1.12)$$

Multi-Cluster Feature Selection (MCFS) is a feature selection method involving **spectral embedding, sparse coefficient vector learning and feature selection based on sparse coefficient vectors** [8].

After calculating the affinity graph W to maintain local geometric structure of the original data space and its diagonal matrix D as well as the graph Laplacian matrix L, **Spectral embedding** is performed by solving the generalized characteristic equation $LF^i = \lambda DF^i$. More specifically, $F = [F^1, \ldots, F^c] \in \mathcal{R}^{n \times c}$ is the embedding for original data and F^i's are the eigenvectors with the smallest eigenvalues. In fact, solving the generalized eigen-problem is equivalent to solving optimization function $\min_{F^T DF = I} tr[F^T LF]$.

Next, **sparse coefficient vectors** are learnt as follows: Given F^i, corresponding sparse coefficient vector w_k can be found through solving the LASSO problem below:

$$\min_{w^i} \left\| F^i - X^T w^i \right\|^2 + \beta \left| w^i \right| \tag{1.13}$$

where $w^i \in R^d, W = [w^1, \ldots, w^c] \in R^{d \times c}$, and $\left| w^i \right| = \sum_{k=1}^d |w_{ki}|$. W is the same as projection matrix in essence.

Finally, **feature selection** is performed. For every feature j, MCFS score for the feature is defined as:

$$MCFS(j) = \max_k \left| w_{jk} \right|, j = 1, \ldots, d, k = 1, \ldots, c \tag{1.14}$$

And the feature with higher MCFS score will be selected.

Nonnegative Discriminative Feature Selection (NDFS) [10] merges spectral embedding and sparse coefficient vector learning in one objective function:

$$\min_{F,W} tr[F^T LF] + \alpha(\left\| X^T W - F \right\|_F^2 + \gamma \left\| W \right\|_{2,1}) \tag{1.15}$$

$$s.t. F^T F = I, F_{ij} \geq 0$$

In NDFS, embedding matrix F is viewed as label indicator. Both nonnegative and orthogonal constraints are applied on matrix F so that it will more resemble ideal label indicator, more capable to supply discriminative messages [9].

Unsupervised Discriminative Feature Selection (UDFS) [5], constructs neighboring set $N_k(x_i)$ for every sample, which contains the sample itself and its k nearest neighbors. And then intraclass deviation of every neighboring set is controlled to be as large as possible while total deviation is limited to take small value. As a result, the feature selected will be equipped with the ability of distinguishing different clusters. The objective function of UDFS is as follows:

$$\min_{W, W^T W = I} \sum_{i=1}^n \{ tr[G_{(i)}{}^T H_{k+1} G_{(i)}] - tr[(S_t^{(i)} + \lambda I)^{-1} S_b^{(i)}] \} + \gamma \left\| W \right\|_{2,1} \tag{1.16}$$

Here $S_t^{(i)}$ represents total scatter matrix of the neighboring set $N_k(x_i)$ of sample i and $S_b^{(i)}$ represents between class scatter matrix. Besides, the first term of (1.16) is added to avoid over-fitting, where $G_{(i)}$ represents the scaled label matrix for samples in $N_k(x_i)$. Since labels is unavailable, there is an assumption in UDFS that a linear classifier W can connect data matrix X and label matrix G, that is, $G = X^T W$.

2   The Proposed Algorithm FSDS

Many algorithms [5-8], [11], [12], complete the task of feature selection by comparing the l_2 norm of each row in transformation matrix $W \in \mathcal{R}^{d \times c}$, which can map the d-dimensional original space to the c-dimensional subspace. And the matrix W is usually gained through optimizing different objective functions. Others [6] may use the graph Laplacian to learn some abstract subspace features, which aim to characterizing original high-dimensional data structure. The abstract subspace features representing compressed information from new low-dimensional space, is similar to the "viewpoints". And these "viewpoints" need to be supported by exact "evidence", that is, some concrete selected features in the original space. Inspired by this, we propose to combine the exact selected features with the subspace features to construct a new interpretive feature set. Our framework is formulated as follows:

- feature selection

$$\min_{W} \alpha D(W) + S(W) + \gamma \|W\|_{2,1}$$
$$s.t. W^T W = I \tag{1.17}$$

For efficient classification and clustering, both similarities among samples in the same classes and differences of samples from different classes are important. Here, D(W) is a clustering criterion of sample differences between different classes, S(W) is another clustering criterion to evaluate sample similarities in the same classes and the $l_{2,1}$-norm of transformation matrix W aims to make it sparse and suitable for feature selection. The orthogonal constraint is imposed to avoid arbitrary scaling and trivial solution of all zeros.

- Subspace feature learning

    In this step, via Laplacian Eigenmap algorithm we learn subspace features $F = [F^1, \ldots, F^c] \in \mathcal{R}^{n \times c}$.

- Feature combination

    Finally, we combine the selected features with the subspace features.

2.1   Exploiting Differences from Global Structure

A satisfactory classifier ought to amplify the difference between different classes and as a result we use the trace of global between class scatter matrix S_b to evaluate the difference. Moreover, it is appropriate to limit the intraclass deviation, which can be viewed as error to some extent. Analogously, the trace of intraclass scatter matrix S_e is used to evaluate intraclass deviation.

$$\max tr[S_b] = \max tr[\tilde{X}GG^T \tilde{X}^T]$$
$$\min tr[S_e] = \min tr[\tilde{X}(I_n - GG^T)\tilde{X}^T]$$
$$\Leftrightarrow \min tr(-\tilde{X}GG^T \tilde{X}^T (\tilde{X}\tilde{X}^T + \lambda I)^{-1}) \tag{1.18}$$

where λ is a parameter and λI is added to ensure the term $(\tilde{X}\tilde{X}^T + \lambda I)$ invertible. Since the proposed algorithm is an unsupervised one, G defined in (1.4) is unknown. We assume there is a linear classifier $W \in \mathcal{R}^{d \times r}$ to transform the original features matrix to the corresponding label matrix, that is, $G = X^T W$. It's worth noting that our framework simultaneously utilizes linear features (feature selected) and nonlinear features (subspace feature), which may improve interpretability. Besides, the orthogonal constraint is added to prevent arbitrary scaling and trivial solution, which can also avoid redundancy in learnt transformation matrix W. In this way, the objective function $\alpha D(W)$ in this subsection can be summarized as

$$\min_{W^T W = I} tr[W^T M^{(1)} W], M^{(1)} = -\alpha \tilde{X}\tilde{X}^T (\tilde{X}\tilde{X}^T + \lambda I)\tilde{X}\tilde{X}^T \tag{1.19}$$

2.2   Exploiting Similarities from Local Perspective

To exploit the similarities among the same classes, we first construct a local set $N_k(centre_i)_{i=1}^c$ for each class center, which can also preserve local structure. Additionally, the k-means algorithm is performed for c center points of c classes. $X_i = [x_{i1}, ..., x_{ik}]$ denotes the local data matrix for i-th class center $G_{(i)} = [G_{i1}, ..., G_{ik}]^T \in \mathcal{R}^{k \times c}$ represents as the corresponding scaled label matrix.

Similar to (1.7), the between class scatter matrix $S_b^{(i)}$ for $N_k(centre_i)$ is defined as

$$S_b^{(i)} = \sum_{j=1}^c n_j^{(i)} (\mu_j^{(i)} - \mu^{(i)})(\mu_j^{(i)} - \mu^{(i)})^T \tag{1.20}$$
$$= \tilde{X}_i G_{(i)} G_{(i)}^T \tilde{X}_i^T$$

where $\tilde{X}_i = X_i H_k$. $\mu^{(i)}$ represents the mean of samples in $N_k(centre_i)$, $\mu_j^{(i)}$ represents the mean of samples in the j-th class, $N_k(centre_i)$. If k is a set small enough, it is appropriate to assume that all samples in $N_k(centre_i)$ belong to the same class. Therefore, $S_b^{(i)}$ is close zero. We then define the selection matrix $S_i \in \{0,1\}^{n \times k}$, each column of which only contains one element being 1 and all others being 0. The first column of S_i is used to record the nearest sample of center i, while the last column is for the farthest away, i.e., $(S_i)_{hg} = 1$ if and only if the h-th sample is the g-th nearest sample of the i-th center. Note that $XS_i = X_i$, $G_{(i)} = S_i^T G = S_i^T X^T W$. Then, we define each term $S_i(W)$ of the whole objective function S(W)

$$
\begin{aligned}
S_i(W) &= tr[S_b^{(i)}] + tr[G_{(i)}{}^T H_k G_{(i)}] \\
&= tr[\tilde{X}_i G_{(i)} G_{(i)}{}^T \tilde{X}_i{}^T + G_{(i)}{}^T H_k G_{(i)}] \\
&= tr[G_{(i)}{}^T \tilde{X}_i{}^T \tilde{X}_i G_{(i)} + G_{(i)}{}^T H_k G_{(i)}] \\
&= tr[G^T S_i \tilde{X}_i{}^T \tilde{X}_i S_i{}^T G + G^T S_i H_k S_i{}^T G] \\
&= tr[W^T XS_i \tilde{X}_i{}^T \tilde{X}_i S_i{}^T X^T W + W^T XS_i H_k S_i{}^T X^T W]
\end{aligned}
\tag{1.21}
$$

Because the learned W and G ought to be suitable for all c classes, we add an $G_{(i)}{}^T H_k G_{(i)}$ to each $S_i(W)$ to avoid overfitting. Finally, we conclude the objective function S(W) in this section.

$$
\min_{W^T W = I} tr[W^T M^{(2)} W], M^{(2)} = \sum_{i=1}^{c} XS_i(\tilde{X}_i{}^T \tilde{X}_i + H_k)S_i{}^T X^T
\tag{1.22}
$$

2.3   Sparseness Control

After adding $l_{2,1}$-norm of the transformation matrix W, the objective function for feature selection (1.17) can be rewritten as

$$
\begin{aligned}
&\min_{W^T W = I} tr(W^T M^{(3)} W) + \gamma \|W\|_{2,1} \\
&M^{(3)} = M^{(1)} + M^{(2)} \\
&= \sum_{i=1}^{c} XS_i(\tilde{X}_i{}^T \tilde{X}_i + H_k)S_i{}^T X^T - \alpha X\tilde{X}^T(\tilde{X}\tilde{X}^T + \lambda I_d)^{-1} \tilde{X}X^T
\end{aligned}
\tag{1.23}
$$

2.4   Subspace Feature Learning and Feature Combination

Construct the affinity graph W [6]:

$$W_{ij} = \begin{cases} \exp(-\dfrac{\|x_i - x_j\|^2}{\sigma^2}), x_i \in N_k(x_j) \ or \ x_j \in N_k(x_i) \\ 0, \ otherwise \end{cases} \quad (1.24)$$

where $N_k(x)$ is the set of k-nearest neighbors of x. D is defined as a $D_{ii} = \sum_{j=1}^{n} W_{ij}$ diagonal matrix whose entries are column(or row) sums of W, that is we can gain the graph Laplacian matrix $L = D - W$. The subspace feature can be computed by solving the following generalized eigen-problem:

$$LF = \lambda DF \quad (1.25)$$

$[F^1, \ldots, F^c] \in \mathcal{R}^{n \times c}$, F^i's are the eigenvectors of (1.25) with the smallest eigenvalues, which are also the solution of $min_{F^T DF = I} \ tr[F^T LF]$.

Finally, we combine the selected features X* with the subspace features F to form a new feature set [X* F].

2.5   Iterative Algorithm of FSDS

The optimization problem in (1.23) involves orthogonal constraint and the $l_{2,1}$-norm, which is not smooth, both increasing the difficulties of optimization. In this subsection, we present an iterative algorithm to solve the optimization problem of FSDS. Similar to [5], Lagrange multiplier is applied and L(W) is defined as:

$$\begin{aligned} L(W) &= tr[W^T M^{(3)} W] + \gamma \|W\|_{2,1} + \lambda(W^T W - I_c) \\ &= tr[W^T M^{(3)} W] + \gamma tr[W^T U W] + \lambda(W^T W - I_c) \\ &= tr[W^T (M^{(3)} + \gamma U) W] + \lambda(W^T W - I_c) \end{aligned} \quad (1.26)$$

where U is a diagonal matrix with $U_{ii} = \dfrac{1}{2\|w_i\|_2}$. Setting $\dfrac{\partial L(W)}{\partial W} = 0$, we have

$$\frac{\partial L(W)}{\partial W} = 2(M^{(3)} + \gamma U)W + 2\lambda W = 0$$
$$\Rightarrow (M^{(3)} + \gamma U)W = -\lambda W = \lambda^* W \quad (1.27)$$

Therefore, we can update W when we fix U through solving the eigen-problem (1.27). Note that $W = [w^1, \ldots, w^c]$, w^i are the eigenvectors with the smallest eigenvalues. In practice, $\|w_i\|_2$ could be close to zero. For this case, we can regularize $2U_{ii} = (w_i^T w_i + \varepsilon)^{-1/2}$, where ε is a small constant to prevent complex solution. When ε is close to zero, $2U_{ii}$ approximates $(w_i^T w_i)^{-1/2}$.

We elaborate the details of FSDS algorithm in Algorithm 1 as follow.

Algorithm 1: The FSDS algorithm

1 $M^{(1)} = -\alpha X \tilde{X}^T (\tilde{X}\tilde{X}^T + \lambda I) \tilde{X} X^T$;

2 perform k-means 20 times to find the c class centre;

3 $M^{(2)} = \sum_{i=1}^{c} X S_i (\tilde{X}_i^T \tilde{X}_i + H_k) S_i^T X^T$;

4 $M^{(3)} = M^{(1)} + M^{(2)}$;

5 **Set** t=0 and initialize $U_0 \in \Re^{d \times d}$ as an identity matrix;

6 **repeat**

7     $P_t = M^{(3)} + \gamma U_t$;

8     $W_t = [w^1, ..., w^c]$,where w¹,...,wᶜ are the eigenvector of P_t with smallest c eigenvalues (c,the class number);

9   Update matrix U_{t+1} as $U_{t-1} = \begin{bmatrix} 1/2\|w_t^1\|_2 & & \\ & ... & \\ & & 1/2\|w_t^d\|_2 \end{bmatrix}$

10   t=t+1;

11 **until** *convergence*

12 Sort all d features according to $\|w_t^i\|_2$ in descending order and select the top r ranked features. The selected feature matrix can be marked as $X^* \in R^{m \times r}$

13 Construct the k-nearest neighbor graph and calculate the graph Laplacian matrix L

14 Gain subspace feature $F = [F^1, ..., F^c] \in R^{m \times c}$ by solving the generalized eigen-problem $LF = \lambda DF$. Then combine W and F to form a new feature set $W = [X^* \; F] \in R^{m \times (r+c)}$

2.6   Convergence Analysis

Next, we prove the convergence of proposed iterative procedure in Algorithm 1.
Theorem 1.1 *The iterative approach in Algorithm 1 (line 6 to line 11) makes the objective function (1.23) monotonically decrease in each iteration.*
Proof. According to Eq. (1.26), Eq. (1.27), we can see that

$$L(W_t, U_{t-1}) \leq L(W_{t-1}, U_{t-1}) \tag{1.28}$$

That is to say,

$$tr[W_t^T(M^{(3)}+\gamma U_{t-1})W_t] \le tr[W_{t-1}^T(M^{(3)}+\gamma U_{t-1})W_{t-1}]$$

$$\Leftrightarrow tr[W_t^T M^{(3)}W_t]+\gamma\sum_{i=1}^{d}\frac{\left\|w_i^t\right\|_2^2}{2\left\|w_i^{t-1}\right\|_2} \le tr[W_{t-1}^T M^{(3)}W_{t-1}]+\gamma\sum_{i=1}^{d}\frac{\left\|w_i^{t-1}\right\|_2^2}{2\left\|w_i^{t-1}\right\|_2}$$

$$\Leftrightarrow tr[W_t^T M^{(3)}W_t]+\gamma\|W_t\|_{2,1}-\gamma(\|W_t\|_{2,1}-\sum_{i=1}^{d}\frac{\left\|w_i^t\right\|_2^2}{2\left\|w_i^{t-1}\right\|_2})$$

$$\le tr[W_{t-1}^T M^{(3)}W_{t-1}]+\gamma\|W_{t-1}\|_{2,1}-\gamma(\|W_{t-1}\|_{2,1}-\sum_{i=1}^{d}\frac{\left\|w_i^{t-1}\right\|_2^2}{2\left\|w_i^{t-1}\right\|_2})$$

$$(1.29)$$

Note that the parameter γ is non-negative. According to Lemmas in (Ref. [2]),

$$\sqrt{a}-\frac{a}{2\sqrt{b}} \le \sqrt{b}-\frac{b}{2\sqrt{b}}, \|W_t\|_{2,1}-\sum_{i=1}^{d}\frac{\left\|w_i^t\right\|_2^2}{2\left\|w_i^{t-1}\right\|_2} \le \|W_{t-1}\|_{2,1}-\sum_{i=1}^{d}\frac{\left\|w_i^{t-1}\right\|_2^2}{2\left\|w_i^{t-1}\right\|_2}$$

Thus, we can prove that:

$$tr[W_t^T M^{(3)}W_t]+\gamma\|W_t\|_{2,1} \le tr[W_{t-1}^T M^{(3)}W_{t-1}]+\gamma\|W_{t-1}\|_{2,1} \qquad (1.30)$$

This is to say,

$$L(W_t,U_t) \le L(W_{t-1},U_{t-1}) \qquad (1.31)$$

which indicates that the objective function value monotonically decreases in each iteration. ☐

2.7   Discussion

In this section, the relationship between the proposed method FSDS and prior work MMC will be discussed. MMC shares similar fashions with the feature selection part of FSDS. MMC is formulated as:

$$W = \arg\max_{W^T W=I} tr[W^T(S_b-S_e)W] \qquad (1.32)$$

And the objective function for feature selection part in FSDS is equivalent to the following formulation:

$$\min_{W^T W=I} \beta tr[W^T(\sum_{i=1}^{c} XS_i \tilde{X}_i^{\ T} \tilde{X}_i S_i^{\ T} X^T)W] + \mu tr[W^T(\sum_{i=1}^{c} XS_i H_k S_i^{\ T} X^T)W]$$
$$- \alpha tr[W^T(X\tilde{X}^T(\tilde{X}\tilde{X}^T + \lambda I_d)^{-1}\tilde{X}X^T)W] + \gamma\|W\|_{2,1}$$

$$(1.33)$$

By setting $\beta = 0, \gamma = 0$, the two algorithms are equivalent under the circumstance that $\tilde{X}\tilde{X}^T$ is invertible and λI_d is removed. In this way, equation (1.33) can be rewritten as:

$$\min_{W^T W=I} \mu tr[W^T(\sum_{i=1}^{c} XS_i H_k S_i^{\ T} X^T)W] - \alpha tr[W^T(X\tilde{X}^T(\tilde{X}\tilde{X}^T)^{-1}\tilde{X}X^T)W]$$

$$\Leftrightarrow \max_{W^T W=I} \alpha tr[W^T(X\tilde{X}^T(\tilde{X}\tilde{X}^T)^{-1}\tilde{X}X^T)W] - \mu tr[W^T(\sum_{i=1}^{c} XS_i H_k S_i^{\ T} X^T)W]$$

$$\Leftrightarrow \max_{W^T W=I} \alpha tr[W^T(XH_n H_n^{\ T} X^T(\tilde{X}\tilde{X}^T)^{-1} XH_n H_n^{\ T} X^T)W] - \mu tr[W^T(\sum_{i=1}^{c} X_i H_k H_k^{\ T} X_i^{\ T})W]$$

$$\Leftrightarrow \max_{W^T W=I} \alpha tr[W^T \tilde{X}\tilde{X}^T(\tilde{X}\tilde{X}^T)^{-1}\tilde{X}\tilde{X}^T W] - \mu tr[W^T(\sum_{i=1}^{c} \tilde{X}_i \tilde{X}_i^{\ T})W]$$

$$\Leftrightarrow \max_{W^T W=I} \alpha tr[W^T \tilde{X}\tilde{X}^T W] - \mu tr[W^T(\sum_{i=1}^{c} \tilde{X}_i \tilde{X}_i^{\ T})W]$$

$$\Leftrightarrow \max_{W^T W=I} tr[W^T S_t W] - \frac{\mu}{\alpha} tr[W^T(\sum_{i=1}^{c} S_t^{(i)})W]$$

$$(1.34)$$

Here, $\sum_{i=1}^{c} S_t^{(i)} = \sum_{i=1}^{c} \sum_{l=1}^{k}(x_l^{(i)} - \bar{\mu}^{(i)})(x_l^{(i)} - \bar{\mu}^{(i)})^T$, k represents the neighbor number of neighboring set $N_k(centre_i)$ of every cluster i and c represents cluster number. Moreover, $x_l^{(i)}$ and $\bar{\mu}^{(i)}$ denote sample vector and mean vector of every neighboring set $N_k(centre_i)$ respectively. As a matter of fact, $\sum_{i=1}^{c} S_t^{(i)}$ is equal to shrunken S_e. More concretely, it just sums the distance between the mean vector and its closest k sample vector of every cluster, but not the total sample vector. Thereby, equation (1.34) can be formulated as:

$$\max_{W^T W=I} tr[W^T S_t W] - \frac{\mu}{\alpha} tr[W^T S_e W] \qquad (1.35)$$

If $\mu/\alpha = 2$, (1.35) can be rewritten as:

$$\max_{W^T W=I} tr[W^T(S_t - 2S_e)W] \Leftrightarrow \max_{W^T W=I} tr[W^T(S_b - S_e)W] \qquad (1.36)$$

Then (1.36) is equivalent to (1.32). To sum up, when $\beta = 0, \gamma = 0$ and $\widetilde{X}\widetilde{X}^T$ is invertible, the feature selection algorithm of FSDS is equivalent to MMC.

Compared with MMC, FSDS takes different value of μ/α into consideration and tries to attain the optimal value. Actually, as the between class deviation is always much larger than the intraclass deviation, a larger weight factor μ/α should be assigned for balance. And it's reasonable that μ/α should be larger than 2. Moreover, FSDS grant β positive value, limiting between class deviation of the same cluster to be small. Thus, the similarities among the same classes can be maintained. Last but not least, FSDS adds regularization term to guarantee the sparseness of transformation matrix W and makes matrix W suitable for feature selection. With the advantages mentioned above, FSDS is expected to gain a better result.

3   The Proposed Algorithm FSDS2

The FSDS is a "two-step" strategy that separates the procedure of subspace feature learning and feature selection in the original space. Inspired by [9], we will discuss about the unified framework of FSDS.

3.1   FSDS2 Framework

The FSDS algorithm can be regarded as solving the following problems in sequence:

- $\arg \min\limits_{W^T W = I} tr(W^T M^{(3)} W) + \gamma \|W\|_{2,1}^T$

- $\arg \min\limits_{F^T F = I} tr(F^T LF)$

- $[X^* \ F]$

The connection of the first two steps can be realized by adding a loss function $\|X^T W - F\|_F^2$. Then the objective function will be formulated as

$$\min_{F,W} tr[F^T LF] + \alpha \|X^T W - F\|_F^2 + \beta tr(W^T M^{(3)} W) + \gamma \|W\|_{2,1} \quad (1.37)$$

$$s.t. F^T F = I, F_{ij} \geq 0$$

Here, we substitute the orthogonal constraint on W with the one on F, which is analogous. It's worth noting that we should regard F as pseudo subspace label but not subspace feature, which may contain more accurate and concise information. Moreover, the nonnegative and orthogonal constraints together on F can not only limit every element F_{ik} range from 0 to 1, but also force only one element in each row of F greater than zero. In this way, F is more like the label.

3.2   Iterative Algorithm of FSDS2

As for the problem of optimization, let us define

$$L(W,F) = tr[F^T LF] + \alpha \left\| X^T W - F \right\|_F^2 + \beta tr[W^T M^{(3)} W] + \gamma \left\| W \right\|_{2,1} \quad (1.38)$$
$$+ \lambda^{(2)} \left\| F^T F - I \right\|_F^2 \; s.t. F_{ij} \geq 0,$$

Similar to FSDS, the optimization problem in (1.24) is also not smooth, thus another iterative algorithm is proposed to solve the problem.

When we fix F and upgrade W. We have

$$\frac{\partial L(F,W)}{\partial W} = 0 = 2\alpha X (X^T W - F) + 2\gamma UW + 2\beta M^{(3)} W \quad (1.39)$$
$$\Rightarrow W = \alpha(\alpha XX^T + \gamma U + \beta M^{(3)})^{-1} XF$$

where U has been discussed above (1.26). Substituting W by (1.39), the objective function (1.38) can be rewritten as

$$\min \; tr[F^T M^{(4)} F] + \lambda^{(2)} \left\| F^T F - I \right\|_F^2 \; s.t. F_{ij} \geq 0, \quad (1.40)$$
$$M^{(4)} = L + \alpha(I_n - \alpha X^T (\alpha XX^T + \gamma U + \beta M^{(3)})^{-1} X)$$

Next, we apply multiplicative updating rules to update F according to [9], [15], [16]. Matrix $\phi = (\phi_{ij})$ is introduced to be Lagrange multiplier for the nonnegative constraint on F. Then we get

$$tr[F^T M^{(4)} F] + \lambda^{(2)} \left\| F^T F - I \right\|_F^2 + tr[\phi F^T] \quad (1.41)$$

Setting its derivative with respect to F to 0 and using the KKT constraints $\phi_{ij} F_{ij} = 0$, we obtain the updating rules:

$$F_{ij} \leftarrow F_{ij} \frac{(\lambda^{(2)} F)_{ij}}{(M^{(4)} F + \lambda FF^T F)_{ij}} \quad (1.42)$$

Then, we normalize F as $(F^T F)_{ii} = 1, i = 1, ..., c$. The detailed algorithm is present below, and the name for the unified approach is FSDS2.

Algorithm 2: The FSDS2 algorithm

---

1 $M^{(1)} = -\alpha X\widetilde{X}^T(\widetilde{X}\widetilde{X}^T + \lambda I)\widetilde{X}X^T$;

2 perform k-means 20 times to find the c class centre;

3 $M^{(2)} = \sum\limits_{i=1}^{c} XS_i(\widetilde{X}_i^T\widetilde{X}_i + H_k)S_i^T X^T$;

4 $M^{(3)} = M^{(1)} + M^{(2)}$;

5 **Set** t=0 and initialize $U_0 \in \Re^{d\times d}$ as an identity matrix and initialize $F_t \in \Re^{n\times c}$

6 **repeat**

7 $M^{(4)^{t+1}} = L + \alpha(I_n - \alpha X^T(\alpha XX^T + \gamma U^t + \beta M^{(3)})^{-1}X)$

8 $F_{ij} \leftarrow F_{ij}\dfrac{(\lambda^{(2)}F)_{ij}}{(M^{(4)}F + \lambda FF^T F)_{ij}}$

9 $W^{t+1} = \alpha(\alpha XX^T + \gamma U^t + \beta M^{(3)})^{-1}XF^t$

10   Update the diagonal matrix U_{t+1} as $U_{t+1} = \begin{bmatrix} 1/2\|w_{t+1}^1\|_2 & & \\ & \cdots & \\ & & 1/2\|w_{t+1}^d\|_2 \end{bmatrix}$

11   t=t+1;

12 **until** *convergence*

13 Sort all d features according to $\|w_t^i\|_2$ in descending order and select the top r ranked features

---

3.3   Convergence Analysis

Next, we prove the convergence of proposed iterative procedure for FSDS2 (Algorithm 2).

Theorem 1.2 *The iterative approach in Algorithm 2 (line 6 to line 12) makes the objective function (1.38) monotonically decrease in each iteration.*

  Proof The problem (1.38) is equal to the minimization problem below:

$$\min_{F^T F=I, F\geq 0, W} \psi(F,W,U) = tr[F^T LF] + \alpha\|X^T W - F\|_F^2 + \beta tr(W^T M^{(3)}W) + \gamma\|W\|_{2,1} \tag{1.43}$$

According to Ref. [12], when U_{t-1}, W_{t-1} is fixed, we can reach

$$\psi(W_{t-1},U_{t-1},F_t) \leq \psi(W_{t-1},U_{t-1},F_{t-1}) \tag{1.44}$$

Besides, when U_{t-1}, F_{t-1} is fixed, (1.39) is the optimal solution of the following problem:

$$\min_W \alpha\|X^T W - F\|_F^2 + \beta tr(W^T M^{(3)}W) + \gamma tr[W^T UW] \tag{1.45}$$

In the following, we will prove that updating W and U simultaneously can decrease the value of function $\Psi(W, U, F)$. For simplicity, we define $g(W) = \alpha\|X^TW - F\|_F^2 + \beta\text{tr}[W^TM^{(3)}W]$. According to (1.43) and (1.45), in the t-th iteration, we have:

$$W^t = \underset{w}{\arg\min}\ g(W) + \gamma tr[W^T U W]$$

$$\Rightarrow g(W_t) + \gamma tr[W_t^T U_{t-1} W_t] \leq g(W_{t-1}) + \gamma tr[W_{t-1}^T U_{t-1} W_{t-1}]$$

$$\Rightarrow g(W_t) + \gamma \sum_{i=1}^d \frac{\left\|w_i^t\right\|_2^2}{\left\|w_i^{t-1}\right\|_2} \leq g(W_{t-1}) + \gamma \sum_{i=1}^d \frac{\left\|w_i^{t-1}\right\|_2^2}{\left\|w_i^{t-1}\right\|_2} \quad (1.46)$$

$$\Rightarrow g(W_t) + \gamma\|W_t\|_{2,1} - \gamma\|W_t\|_{2,1} - \sum_{i=1}^d \frac{\left\|w_i^t\right\|_2^2}{\left\|w_i^{t-1}\right\|_2}) \leq g(W_{t-1}) + \gamma\|W_{t-1}\|_{2,1} - \gamma\|W_{t-1}\|_{2,1} - \sum_{i=1}^d \frac{\left\|w_i^{t-1}\right\|_2^2}{\left\|w_i^{t-1}\right\|_2})$$

Here the parameter γ is non-negative. According to Lemmas in (Ref. [2]), $\sqrt{a} - \frac{a}{2\sqrt{b}} \leq \sqrt{b} - \frac{b}{2\sqrt{b}}$, we can analogously prove:

$$\sum_{i=1}^d (\left\|w_i^t\right\|_2 - \frac{\left\|w_i^t\right\|_2^2}{\left\|w_i^{t-1}\right\|_2}) \leq \sum_{i=1}^d (\left\|w_i^{t-1}\right\|_2 - \frac{\left\|w_i^{t-1}\right\|_2^2}{\left\|w_i^{t-1}\right\|_2})$$

$$\Leftrightarrow \left\|W_t\right\|_{2,1} - \sum_{i=1}^d \frac{\left\|w_i^t\right\|_2^2}{\left\|w_i^{t-1}\right\|_2} \leq \left\|W_{t-1}\right\|_{2,1} - \sum_{i=1}^d \frac{\left\|w_i^{t-1}\right\|_2^2}{\left\|w_i^{t-1}\right\|_2} \quad (1.47)$$

Therefore, the inequality below is proved:

$$g(W_t) + \gamma\|W_t\|_{2,1} \leq g(W_{t-1}) + \gamma\|W_{t-1}\|_{2,1} \quad (1.48)$$

That is,

$$\psi(W_t, U_t, F_t) \leq \psi(W_{t-1}, U_{t-1}, F_t) \quad (1.49)$$

Based on (1.44) and (1.49), we can finally conclude that:

$$\psi(W_t, U_t, F_t) \leq \psi(W_t, U_t, F_{t-1}) \leq \psi(W_{t-1}, U_{t-1}, F_{t-1}) \quad (1.50)$$

which indicates that the objective function value monotonically decreases in each iteration.

3.4   Discussion

In this section, we will discuss about the connection between the proposed algorithm FSDS2 and prior algorithms NDFS, MCFS.

NDFS, Nonnegative Discriminative Feature Selection. As is mentioned in 1.2, NDFS can be formulated as the following form.

$$\min_{F,W} \ tr[F^T LF] + \alpha(\|X^TW - F\|_F^2 + \gamma\|W\|_{2,1})$$

$$s.t. F^T F = I, F_{ij} \geq 0$$

(1.51)

And the objective function for FSDS2 is as follows:

$$\min_{F,W} \ tr[F^T LF] + \alpha\|X^TW - F\|_F^2 + \beta tr(W^T M^{(3)}W) + \gamma\|W\|_{2,1}$$

$$s.t. F^T F = I, F_{ij} \geq 0$$

(1.52)

It is obvious that when $\beta = 0$ in FSDS2 minimization problem (1.52), the two algorithms are the same. Hence, NDFS is a special case of the proposed FSDS2. For the reason that FSDS2 exploits both similarities among neighboring samples and differences among different classes in the third term of objective function (1.52), FSDS2 tends to be more comprehensive and discriminative than NDFS.

MCFS, Multi-Cluster Feature Selection, achieve feature selection via the following optimization functions:

$$\min_{F^T DF=I} \ tr[F^T LF]$$

(1.53)

$$\min_{w^i} \ \|F^i - X^T w^i\|^2 + \beta|w^i| \quad \text{for every } F^i$$

$$\Leftrightarrow \min_{W=[w^1,\ldots,w^c]} \ \sum_{i=1}^c (\|F^i - X^T w^i\|^2 + \beta|w^i|)$$

(1.54)

where $w^i \in R^d$, $W = [w^1, \ldots, w^c] \in R^{d \times c}$, and $|w^i| = \sum_{k=1}^d |w_{ki}|$.

If squared l_2 norm is applied to the regularization term of MCFS and the squared $l_{2,2}$ norm is applied to the one of FSDS2, then MCFS will be a special case of FSDS2.

First, set $\beta = 0$ and remove the nonnegative constraint of F in FSDS2. (1.52) will be rewritten as:

$$\min_{F,W} tr[F^T LF] + \alpha \left\| X^T W - F \right\|_F^2 + \gamma \|W\|_{2,2}^2$$

$$s.t. F^T F = I$$

(1.55)

More concretely, $l_{2,2}$ norm of W can be formulated as $\|W\|_{2,2} = (\sum_{k=1}^d (\sum_{i=1}^c w_{ki}^2))^{1/2}$. And the squared $l_{2,2}$ norm can be written as $\|W\|_{2,2}^2 = \sum_{k=1}^d (\sum_{i=1}^c w_{ki}^2)$.

Then, set $\alpha \to 0, \gamma \to 0$ and (1.55) will be divided into two sub-problems (1.56), (1.57).

$$\min_{F, F^T F=I} tr[F^T LF]$$

(1.56)

$$\min_{W} \alpha \left\| X^T W - F \right\|_F^2 + \gamma \|W\|_{2,2}^2$$

$$\Leftrightarrow \min_{W} \sum_{i=1}^c \left(\left\| X^T w^i - F^i \right\|_2^2 + \frac{\gamma}{\alpha} \left\| w^i \right\|_2^2 \right)$$

(1.57)

Here $w^i \in R^d, W = [w^1, \ldots, w^c] \in R^{d \times c}, \|w^i\|_2^2 = \sum_{k=1}^d w_{ki}^2$. Actually, the squared l_2 norm on regularization term of MCFS (1.54) can be elaborated as $|w^i|^2 = \sum_{k=1}^d |w_{ki}|^2$. Thus, (1.56), (1.57) is equal to (1.53), (1.54).

Similarly, FSDS2 take into account both similarities and differences among samples via the third term of objective function (1.52) while this work is not involved in MCFS. Moreover, for FSDS2, label indicator matrix F and transformation matrix W are learnt together in one optimization function so that the priority of different F^i's is consider when label indicator matrix F instructs the learning of transformation matrix W. However, for MCFS, different F^i's guide the learning of corresponding transformation vector w^i independently without considering the competitive relationship of different F^i's. As a result, FSDS2 seems to be more reasonable and capable of selecting effective features.

4   Experiments

In this section, we conduct several experiments to evaluate the performance of FSDS and FSDS2. We test the performance of the proposed algorithms in terms of clustering results.

4.1 Datasets

The experiments are conducted on 5 public datasets, including 3 face image datasets, Yaleface, JAFFE [17] and UMIST [16], one hand-written digit image dataset, USPS [16] and one object image dataset, coil20 [18]. Table 1 summarizes the details of these datasets mentioned above.

Table 1. Dataset description

Dataset	#of Samples	#of Features	#of Class
Yaleface	165	1024	15
JAFFE	213	676	10
UMIST	575	644	20
USPS	400	256	10
coil20	1440	1024	20

4.2 Experiment Settings

We compare the performance of FSDS and FSDS2 with the following unsupervised feature selection algorithm.

- **Baseline:** All original features are adopted.
- **MaxVar:** Features with maximum variance are selected.
- **LPP** [7]: Subspace features gain by Laplacian Eigenmap algorithm.
- **LS** [6]: Features consistent with Gaussian Laplacian matrix are selected.
- **MCFS** [8]: Features are selected through spectral analysis and sparse regression problem.
- **UDFS** [5]: Features are selected based on local discriminative feature analysis and $l_{2,1}$-norm minimization.
- **NDFS** [10]: Features are selected by nonnegative spectral analysis and linear regression with $l_{2,1}$-norm regularization.
- **CGSSL** [12]: Features are selected by also considering the original features when predicting the pseudo label based on NDFS.
- **FSDS:** Features are selected by the proposed method Feature Selection with Differences and Similarities (algorithm 1).
- **FSDS2:** Features are selected by the joint framework of FSDS (algorithm 2).

For all algorithm except Baseline, MaxVar, we set k, the size of neighborhoods, to be 5 for all the datasets. For NDFS, CGSSL, FSDS2, we fix $\lambda^{(2)} = 10^8$ to guarantee the orthogonality of F and fix $\lambda = 10$ to ensure the term $(\tilde{X}\tilde{X}^T + \lambda I)$ invertible. To fairly compare the performance of all algorithms, we tune the parameters for all approaches by a "grid search" strategy from $\{10^{-6}, 10^{-4}, ..., 10^{6}\}$. The numbers of selected features are set as $\{50,100,150,200,250,300\}$ for all of the datasets except for USPS, whose total feature number is 256. We set the selected feature number of USPS as $\{50,80,110,140,170,200\}$. We report the best results of all these 10 algorithms with different parameters.

After feature selection through different algorithm, k-means clustering is conducted based on the selected features. Besides, k-means is also performed in FSDS and FSDS2 feature selection procedure. Since the results of the k-means is greatly affected by its initialization, we repeat it 20 times with random initialization every time we use it. We report the average results with standard deviation (std).

We apply two evaluation metrics, Accuracy (ACC) and Normalized Mutual Information (NMI), to evaluate the clustering performance. The larger ACC and NMI are, the better the performance is. ACC is defined as follows:

$$ACC = \frac{\sum_{i=1}^{n} \delta(p_i, map(q_i))}{n} \qquad (1.58)$$

where $(a, b) = 1, if\ a = b, \delta(a, b) = 0\ otherwise$, and q_i is the clustering label and p_i is the ground-truth label of x_i. *map* is a kind of permutation function that maps each cluster index to the best ground-truth labels using the Kuhn-Munkres algorithm. Mutual Information (MI) and Normalized Mutual Information (NMI) are defined as follows:

$$MI(P,Q) = \sum_{c_i \in P} \sum_{c_j' \in Q} p(c_i, c_j') \log_2 \frac{p(c_i, c_j')}{p(c_i)p(c_j')}$$

$$NMI(P,Q) = \frac{MI(P,Q)}{\max(H(P), H(q))} \qquad (1.59)$$

where p(c) is the probability that a sample randomly selected from the dataset belongs to class c. P represents clustering label set whereas Q represents groud truth label set. And H(c) is the entropy of c. $H(c) = -\sum_{c_i \in c} p(c_i) \log_2 p(c_i)$.

4.3   Experimental Results

We summarize the clustering results of different methods on the 5 datasets in Table 2.

Table 2. Clustering Results of Different Feature Selection Algorithm

Dataset	ACC±std(%)									
	Baseline	MaxVar	LS	LPP	MCFS	UDFS	NDFS	CGSSL	FSDS	FSDS2
Yale	46.7 ±3.3	40.6 ±2.9	50.5 ±3.5	45.5 ±3.2	50.3 ±5.1	49.1 ±3.8	45.5 ±3.0	-	50.9 ±4.4	50.9 ±3.7
JAFFE	72.5 ±9.2	67.3 ±5.8	74.0 ±7.6	69.7 ±2.4	78.8 ±9.1	76.7 ±7.1	81.2 ±8.1	82.3 ±7.5	95.8 ±8.0	95.3 ±9.1
USPS	62.6 ±5.3	63.8 ±4.3	64.9 ±5.1	64.0 ±1.8	64.4 ±3.1	66.2 ±4.7	67.3 ±4.7	68.3 ±4.5	78.8 ±5.8	78.8 ±5.5
coil20	59.0 ±5.7	58.4 ±4.0	57.3 ±3.0	50.8 ±5.0	61.7 ±4.3	62.9 ±2.6	63.8 ±2.8	-	72.6 ±4.7	72.9 ±4.3
UMIST	41.8 ±2.7	45.8 ±2.8	45.9 ±2.9	56.5 ±2.4	46.3 ±3.6	48.6 ±3.7	51.3 ±3.9	53.4 ±3.1	58.8 ±4.2	60.5 ±3.8
	NMI±std(%)									
Yale	48.3 ±2.8	46.6 ±2.0	54.3 ±2.6	50.6 ±2.9	57.7 ±3.6	53.9 ±3.4	49.4 ±2.7	-	55.2 ±3.4	57.0 ±3.1
JAFFE	80.0 ±5.7	70.3 ±4.2	79.4 ±7.0	80.8 ±3.3	83.4 ±5.0	82.3 ±6.5	86.3 ±7.1	87.5 ±5.1	93.8 ±5.5	93.5 ±6.9
USPS	56.9 ±3.1	58.1 ±2.7	58.7 ±3.0	62.4 ±1.5	59.3 ±2.9	60.1 ±4.3	61.3 ±2.5	62.0 ±2.8	68.1 ±3.4	68.1 ±3.3
coil20	72.9 ±2.8	70.5 ±0.9	70.4 ±1.1	61.9 ±3.4	74.7 ±2.3	75.9 ±1.1	77.1 ±1.8	-	80.6 ±2.1	80.2 ±3.1
UMIST	62.3 ±2.3	63.5 ±1.5	63.9 ±1.8	74.2 ±3.0	66.7 ±1.9	67.3 ±3.0	69.7 ±2.3	70.9 ±2.2	73.3 ±2.8	73.1 ±2.7

Through observing this table, we can gain the following conclusion. First, it is necessary to apply feature selection not only because feature selection can increase efficiency and save experiment time, but also because it can reduce useless noise in datasets and improve clustering performance, which can be supported by the fact that most feature selection algorithm outperform Baseline. Second, it places great importance to consider local structure in feature selection, which is consistent with the fact that other algorithms get better performance than MaxVar, the only one algorithm without preserving local structure. Third, the proposed Algorithm FSDS and FSDS2 achieve best performance in most datasets by learning the linear classifier W that focuses on both sample differences and sample similarities. Other algorithms just take either of sample differences or similarities into account. For example, LS, MCFS, NDFS, CGSSL preserve the manifold structure by mapping the neighboring data point close to each other in the mapping subspace, which is a emphasis on sample similarities. And UDFS learn the classifier W that can recognize the sample difference. It may be more comprehensive to consider both of differences and similarities. Hence, FSDS and FSDS2 are proved to be capable of more efficient feature selection.

Moreover, FSDS and FSDS2 achieve similar ACC and NMI, which may be owing to the following reasons. We can easily find in Table 3 that the information of subspace features and the information of original features remain independent and complete after FSDS algorithm. For FSDS2, according to (1.23), the information of subspace features and original features fuses together and they affect each other. Both these two solutions have their advantages, thus resulting in the similar results.

Table 3. Clustering Results of FSDS with Subspace Feature Learning and without

Dataset	*ACC±std(%)*	
	with Subspace Feature	*without Subspace Feature*
Yale	50.9±4.4	47.3±4.3
JAFFE	95.8±8.0	92.5±7.8
USPS	78.8±5.8	76.8±4.3
coil20	72.6±4.7	65.5±4.0
UMIST	58.8±4.2	55.5±3.8

4.4   Convergence Study

In the previous section, we have proved the convergence of the proposed alternative optimization algorithm for FSDS and it is analogy for FSDS2. Our algorithms converge within 50 iterations for all the data sets, which indicates that the proposed update algorithms are effective and converge fast, especially the one for FSDS2. Moreover, the results of convergence are shown in Fig. 1.

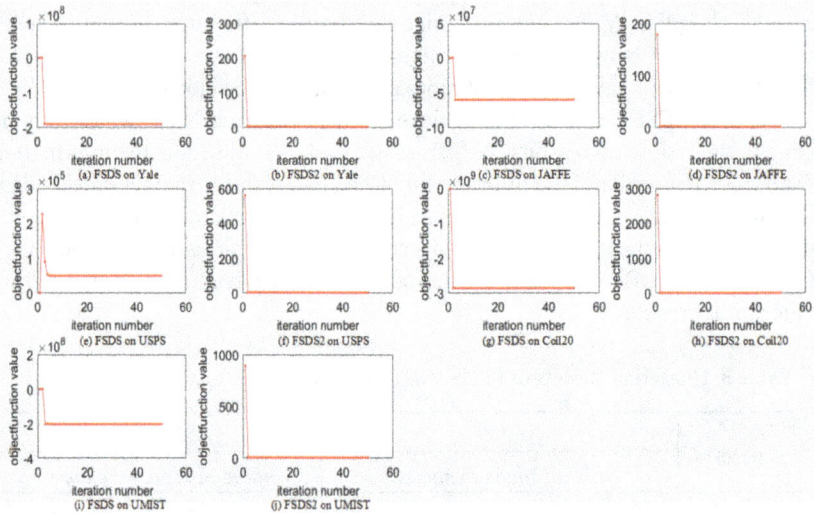

Fig. 1. Convergence Curve of the Proposed Optimization Algorithm for FSDS and FSDS2

4.5   Sensitivity Analysis

For FSDS method, there are three parameters α, γ, f ($feature\ number$) in need of setting in advance. And FSDS2 algorithm requires five parameters to be tuned, i.e. $\alpha, \gamma, \beta, \alpha^s$ ($parameter\ in\ M^{(3)}$), f ($feature\ number$). For the results reported in the first subsection, we do not discuss about number of selected feature. Thus, we will investigate the sensitivity of α, γ for FSDS. As for FSDS2, we observe that the parameters γ, β have more effect on the performance than others in our experiment, which will be studied in the following.

All the parameters mentioned above are tuned from $\{10^{-6}, 10^{-4}, ..., 10^6\}$. The results over the five data sets are shown in Fig. 2 and Fig. 3. In conclusion, there seems no analogous rule to determine the optimal parameters on all condition due to the varying performance with respect to different data sets. In terms of FSDS, firstly, the parameter γ controls the row sparsity of the feature selection matrix W. when γ is set to be too small, noisy or redundant features can't be spotted and removed. Instead, not only useless features but also informative features will be discarded if γ is too large. These are consistent with the observation in Fig. 2 that extremely large or extremely small γ has negative influence on the performance. Secondly, the α parameter balances the importance between similarity preserving and difference exploiting. It can be concluded from Fig. 3 that we always attain the best performance while α is in the middle interval, demonstrating that similarity preserving and difference exploiting are of equal importance.

With regard to FSDS2, on one hand, the analysis for parameters γ similar to the one mentioned above. On the other hand, the parameter β is the trade-off parameter between feature selection according to (1.23) and subspace feature learning. It is observed from Fig. 3 that parameter β, neither too small nor too large, frequently accompanies optimal results. This indicates that the information in the original feature space is as essential as the information in abstract subspace.

Next, we evaluate the performance of both proposed methods and compared methods with different value of f (*feature number*). The experimental results are present in Fig. 4, from which we can observe the following conclusions. First, the performance is comparatively sensitive to the number of selected features. Second, the proposed algorithms FSDS and FSDS2 can always reach the best results with some small values of selected features. It can be verified that the proposed algorithms not only comprehensively exploit the differences and similarities among the samples, but they also synthetically investigate the original feature space and the underlying manifold space, thus resulting in efficient feature selection.

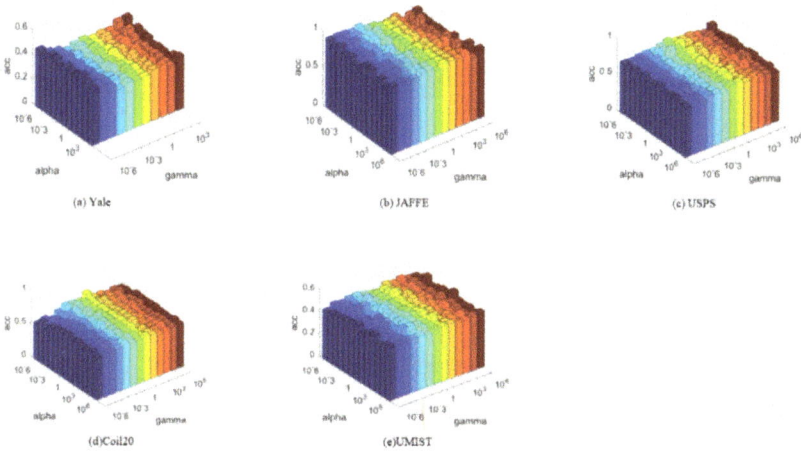

Fig. 2. Parameter Sensitivity for FSDS

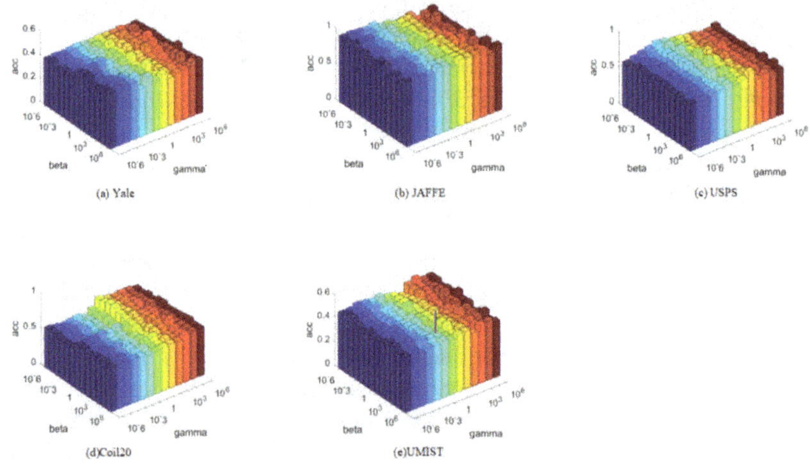

(a) Yale

(b) JAFFE

(c) USPS

(d)Coil20

(e)UMIST

Fig. 3. Parameter Sensitivity for FSDS2

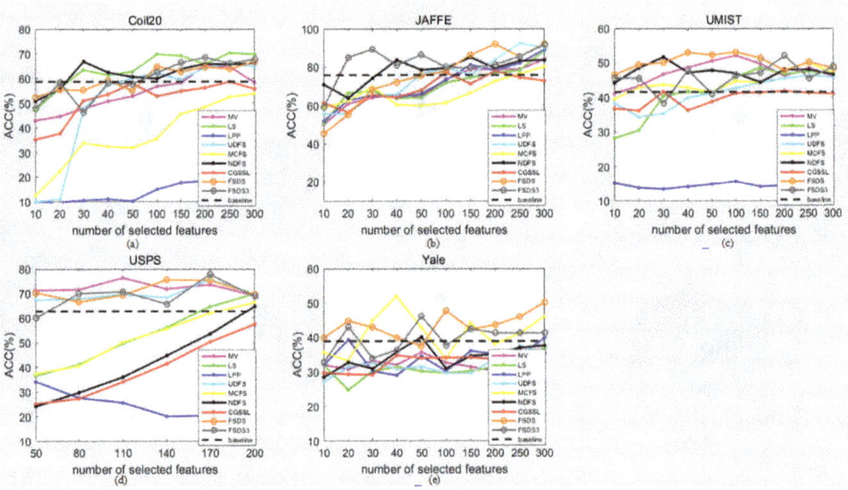

Fig. 4. The clustering performance in terms of ACC (%) with respect to the number of selected features on all the five datasets

5   Conclusions

In this chapter, we propose two feature selection approaches FSDS and FSDS2, which combine the feature selection procedure and spectral subspace feature learning. Utilizing both original feature and subspace feature, the subsequent clustering methods gains better performance. Besides, our methods also jointly exploit sample differences and sample similarities. Extensive experiments on different datasets have validated the effectiveness of the proposed methods.

References

1. L. Wolf and A. Shashua. (2005). Feature selection for unsupervised and supervised inference: The emergence of sparsity in a weightbased approach, *J. Mach. Learn. Res.*, 6, pp. 1855–1887.
2. F. Nie, H. Huang, X. Cai, and C. Ding. (2010). Efficient and robust feature selection via joint l2,1-norms minimization, *Proc. of the Advances in Neural Information Processing Systems*, NIPS: 1813–1821.
3. Z. Zhao and H. Liu. (2007). Semi-supervised feature selection via spectral analysis, *Proc. of the SIAM International Conference on Data Mining*, SDM: 641–646.
4. Xu. Z, King. I, Lyu. R T, & Jin. R. (2010). Discriminative semi-supervised feature selection via manifold regularization, IEEE Transactions on Neural Networks, 21, pp. 1033–1047.
5. Y. Yang, H. T. Shen, Z. Ma, Z. Huang, and X. Zhou. (2011). l2,1-norm regularized discriminative feature selection for unsupervised learning, *Proc. of the Twenty-Second international joint conference on Artificial Intelligence*, IJCAI: 1589–1594.
6. X. He, D. Cai, and P. Niyogi. (2005). Laplacian score for feature selection, *Proc. of the Advances in Neural Information Processing Systems*, NIPS: 507–514.
7. X. He, and P. Niyogi. (2003). Locality Preserving Projections, *Proc. of the Advances in Neural Information Processing Systems*, NIPS: 186–197.
8. D. Cai, C. Zhang, X. He. (2010). Unsupervised Feature Selection for Multi-Cluster Data, *Proc. ACM SIGKDD Conf. on Knowledge Discovery and Data Mining (KDD'10)*: 333–342.
9. C. Hou, F. Nie, X. Li, D. Yi, and Y. Wu. (2014). Joint Embedding Learning and Sparse Regression: A framework for Unsupervised Feature Selection, IEEE Transactions on Cybernetics, 44, pp. 793–804.
10. Z. Li, Y. Yang, J. Liu, X. Zhou, and H. Lu. (2012). Unsupervised feature selection using nonnegative spectral analysis, *Proc. of the Twenty-Sixth AAAI Conference on Artificial Intelligence*, AAAI: 1026–1032.
11. J. Shi and J. Malik. (2000). Normalized cuts and image segmentation, IEEE Transactions on Pattern Analysis and Machine Intelligence, 22, pp. 888–905.
12. Z. Li, Y. Yang, J. Liu, X. Zhou, and H. Lu. (2014). Clustering-Guided Sparse Structural Learning for Unsupervised Feature Selection, IEEE Transactions on Knowledge and Data Engineering, 26, pp. 2138–2150.
13. Z. Li, J. Liu, J. Tang, and H. Lu. (2015). Robust Structured Subspace Learning for Data Representation, IEEE Transactions on Pattern Analysis and Machine Intelligence, 37, pp. 2085–2098.
14. Z. Li, J. Tan. (2015). Unsupervised Feature Selection via Nonnegative Spectral Analysis and Redundancy Control, IEEE Transactions on Image Processing, 24, pp. 5343–5355.

15. D. Lee, and H. Seung. (1999). Learning the parts of objects by nonnegative matrix factorization, Nature, 401, pp. 788–791.
16. D. Lee, and H. Seung. (2000). Algorithms for nonnegative matrix factorization, *Proc. of the Advances in Neural Information Processing Systems*, NIPS: 535–541.
17. Data for MATLAB Hackers. Available: http://cs.nyu.edu/~roweis/data.html.
18. M. J. Lyons, J. Budynek, and S. Akamatsu. (1999). Automatic classification of single facial images, IEEE Transactions on Pattern Analysis and Machine Intelligence, 21, pp. 1357–1362.
19. Columbia object image library (COIL-20). Available: http://www.cs.columbia.edu/CAVE/software/softlib/coil-20.php
20. M. Luo, X. Chang, L. Nie, Y. Yang, A. G. Hauptmann and Q. Zheng. (2018). An Adaptive Semisupervised Feature Analysis for Video Semantic Recognition, IEEE Transactions on Cybernetics, 48, pp. 648–660.
21. Z. Zhao, X. He, D. Cai, L. Zhang, W. Ng and Y. Zhuang. (2016). Graph Regularized Feature Selection with Data Reconstruction, IEEE Transactions on Knowledge and Data Engineering, 28, pp. 689–700.
22. Z. Cai and W. Zhu. (2018). Feature selection for multi-label classification using neighborhood preservation, IEEE/CAA Journal of Automatica Sinica, 5, pp. 320–330.
23. N. Zhou, Y. Xu, H. Cheng, Z. Yuan and B. Chen. (2019). Maximum Correntropy Criterion-Based Sparse Subspace Learning for Unsupervised Feature Selection, IEEE Transactions on Circuits and Systems for Video Technology, 29, pp. 404–417.
24. Y. Jia, F.-P. Nie, C. Zhang. (2009). Trace ratio problem revisited, IEEE Transactions on Neural Networks, 20, pp. 729–735.
25. Y.-F. Guo, S.-J. Li, J.-Y. Yang, et al. (2003). A generalized Foley-Sammon transform based on generalized fisher discriminant criterion and its application to face recognition, Pattern Recognition Letters, 24, pp. 147–158.
26. H. Yu and J. Yang. (2001). A direct LDA algorithm for high-dimensional data with application to face, recognition, Pattern Recognition, 34, pp. 2067–2070.

Index

blood cell classification   27, 31, 32, 37

blood smear analysis   23, 24, 27

breast lesion detection   1, 2

character amnesia   98, 99, 114

classification   3, 4, 6, 19, 26, 27, 29–37, 39, 49, 57, 61, 67, 68, 75, 109, 119–124, 127, 129, 141, 145, 146, 153, 164, 168, 179, 193–197, 199, 201, 205–208, 214, 220, 228, 232, 237

computer aided diagnosis   23, 63

Convolutional Neural Network (CNN)   3–6, 19, 28–30, 32–38, 48, 50, 52–58, 68, 100, 105, 109, 119–121, 123–125, 128, 168–170, 172–174, 197, 213–215, 218, 220, 222–224, 226

credit scoring   145–147, 152, 158

Crisi Wartegg System   136, 139, 140, 142

data augmentation   61, 63, 69, 71, 72, 74, 75

deep CNN   3, 5, 6, 69, 120

deep learning   1, 2, 4, 5, 23, 24, 33, 40, 48, 49, 57, 61, 63, 65–69, 75, 119, 120, 123, 164, 165, 169, 193, 194, 197, 205, 214, 215

feature choice   161, 162, 179

feature map   4, 61, 105

feature selection   35, 177, 178, 196, 231, 232, 234–237, 239, 242–244, 248, 250–256

features by learning   161

functional intelligence   133

gender classification   119, 121, 123, 124

generative adversarial network   47, 49, 50, 52, 54, 57, 58, 69

hand-crafted   47–50, 57, 67, 165, 168, 169, 174, 178

handwriting   77–94, 97–104, 109–114, 120–123, 125, 127–129, 218

handwriting analysis   93, 94, 99, 119, 120

handwriting difficulty   97–101, 103, 104, 109, 111–114

image analysis   vi, 61, 63, 69, 161–163, 166, 169–173, 176–178

image database   26

leukemia   23, 26, 27, 34–36, 38, 39

license plates   183–191

machine learning and prediction   228

Magnetic Resonance Imaging (MRI)   2, 3, 6, 7, 10, 19

malaria   27–30, 36

medical CADe   1, 2

natural language processing   193,
        194, 200, 208
neural networks   2, 23, 28, 31, 37,
        48, 168, 193–196, 198–200,
        207
news classification   195

palm-vein recognition   47
personality   77, 79–81, 83, 84, 87–
        91, 93, 94, 138, 140, 141
psychology   120

regression   3–5, 8, 9, 15, 16, 19,
        100, 102, 107, 109, 112, 113,
        146, 196, 197, 250
reinforcement learning   97, 99,
        100, 102, 104, 107, 108, 112–
        114

retina fundus photographs   61, 62,
        70
retinal image   61, 63, 65, 72–75
row sparsity   254
signal processing on graph   145,
        152
social media   197, 213, 214
Spanish criminal news   193, 195,
        201, 202, 205, 207
spectral learning   231

traits   48, 77, 79–81, 84, 87, 88, 94,
        120
transfer-learning   3, 4, 6, 119, 120,
        125, 128, 168, 200

U-Net   38, 47, 49, 50, 52, 58, 172